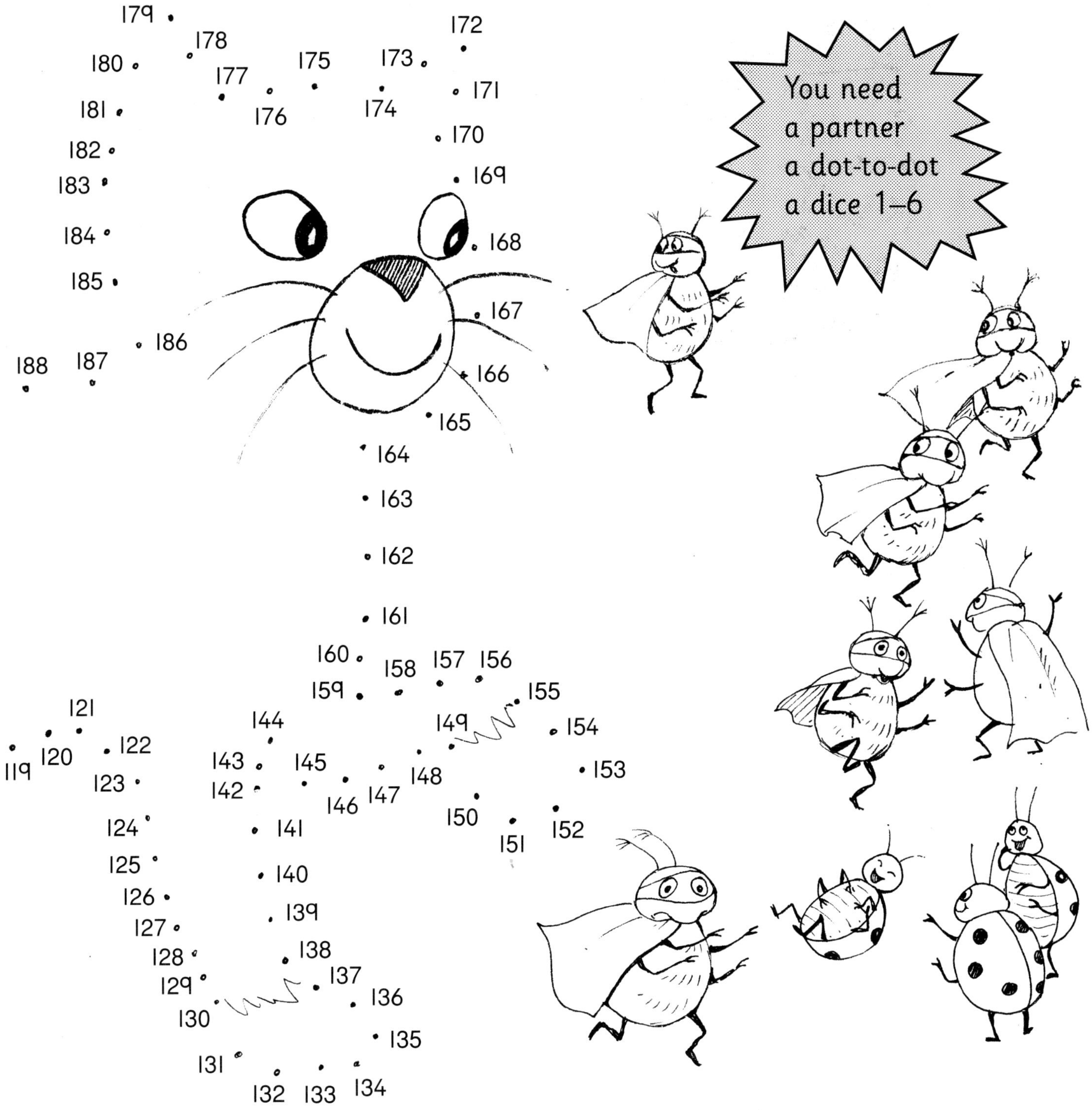

You need
a partner
a dot-to-dot
a dice 1–6

179
178
180
177
175
173
172
181
176
174
171
182
170
183
169
184
168
185
167
186
166
188 187
165
164
163
162
161
160 158 157 156
159 155
144 149 154
121
143 145 148 153
120 122
119 123 142 146 147 150 152
124 141 151
125 140
126 139
127 138
128 137 136
129 130 135
131 134
132 133

Name . . . . **Class** . . . .

Rules

- Take turns to roll the dice.
- Join up that number of dots. (If you roll a 2, join the next 2 dots).
- The winner is the player who completes the picture.

Start

Kit to the rescue!

Name . . . . **Class** . . . .

Count in twos:

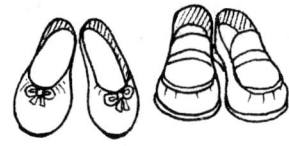  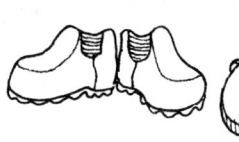 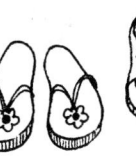  

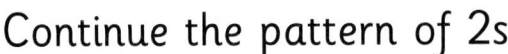 shoes altogether

Continue the pattern of 2s

10　12　14　☐　☐　☐　☐　☐　☐　☐

22　24　☐　28　☐　☐　☐　☐　☐　40

Count back in 2s

22　20　18　☐　☐　☐　☐　☐　☐　☐

82　80　78　☐　☐　☐　☐　☐　☐　☐

Count in 5s

 brushes altogether

5　10　☐　☐　☐　☐　☐　40　☐　☐

Count back in 5s

50　45　40　☐　30　☐　☐　☐　☐　☐

2	Before you start IP 1	● Can count in steps of 2 or 5 from any small number. Notes/date:

3

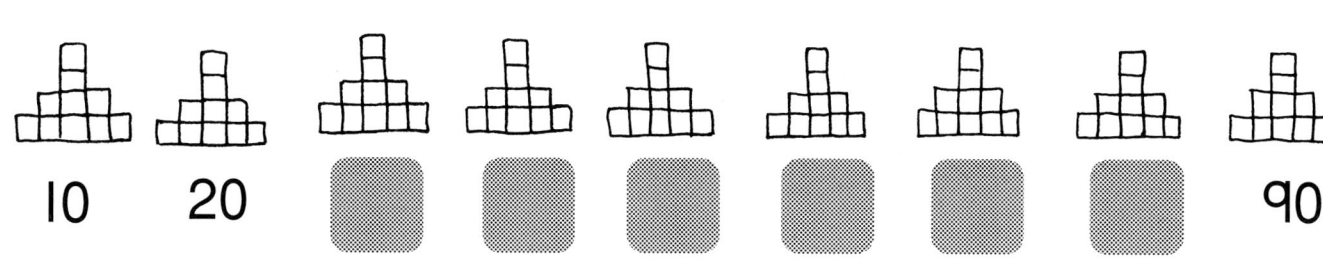

10   20   ▨   ▨   ▨   ▨   ▨   ▨   90

Continue the pattern:

60   70   ▨   ▨   100   110   ▨   ▨   ▨

Count back in tens:

160   150   ▨   ▨   120   ▨   ▨   90   ▨

Count forward in ones:

98   99   100   101   102   ▨   ▨   ▨   ▨

198   199   200   ▨   ▨   ▨   ▨   ▨

Count back in ones:

105   104   103   ▨   ▨   ▨   ▨   ▨

Put the numbers on the pegs:

● Can count on or back in steps of one or ten from any number.
Notes/date:

Before you start
IP 1

3

Name . . . . Class . . . .

Help Nevis count in 10s:

30 40 ▢ ▢ ▢ ▢ ▢ 100 110 ▢ ▢ 140

and back:

200 190 180 ▢ ▢ ▢ ▢ ▢ ▢

110 100 ▢ ▢ ▢ ▢ 50 ▢ ▢ ▢ ▢

Zero!

Count in 5s

205 210 215 ▢ ▢ ▢ ▢

Write the next page number:

| 38 | 39 | | 47 | | | 69 | | | 84 | | | 100 | |

| 52 | | | 114 | | | 119 | | | 123 | | | 109 | |

Write the numbers:

Fifty-six ▢   Eighty-eight ▢   One hundred and two ▢

| 4 | Before you start IP 1 | ● Can count on or back in steps of 1, 5 or 10 from any number. Notes/date: |

Name . . . . Class . . . .

Count in 3s:

There are children altogether

Continue the pattern of 3s:

0   3   6   9

1   4   7    16

Count back in 3s:

36 33

Count in 4s:

There are circles in the palettes altogether

0   4   8     24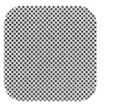

36 40    60

Count back in 4s:

28 24     0

● Can count in steps of 3 or 4.
Notes/date:

Before you start
IP 1

5

6

Name . . . . **Class** . . . .

Circle the multiples of 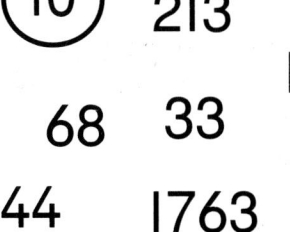 213   20   75

87   30   90   1000 102

73   40   68   33

100   200   60   29   70   61   80

856   44   1763

Make up more beetle sentences here

$6 + 4 =$

$4 + \boxed{} = 10$

$10 - 6 =$

$10 - 4 =$

Bandit Beetle whispers to 10

Circle the multiples of (5)

56   40   73   20   10   25   45

35   60   53

8   5   75   12

30   55   100   96   142

15   95   24   18   31   48

6	Before you start IP 1	● Knows addition and subtraction facts for numbers to 10. ● Begins to recognise multiples of 5 or 10. Notes/date:

Name . . . . **Class** . . . .

I wish I had twice as many marbles, McMagic

Then I'll **double** them for you Nevis!

ZAP!

Doubling makes numbers twice as big.

Double these numbers:

double ↘	3	2	7	4	8	10	9	12	6	11	20	50	100

Double the length of these lines:

5 cm → 10 cm

3 cm

6 cm

4 cm

Double the creatures:

3 beetles

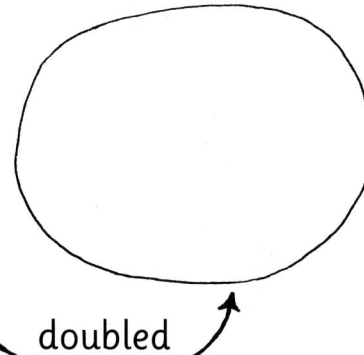

doubled

2 cats

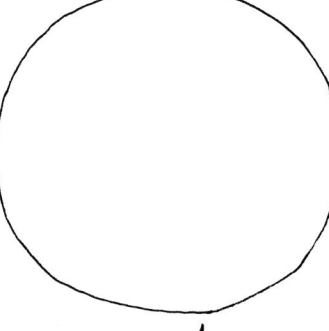

doubled

● Knows doubles of numbers to 10 and multiples of 10.
Notes/date:

Before you start
IP 2

7

Name . . . . **Class** . . . .

Double 9 is 18.
Half of 18 is 9.

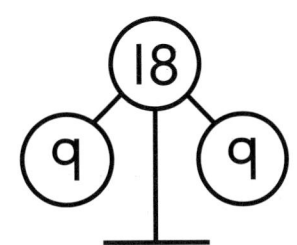

Fill in these balances:

  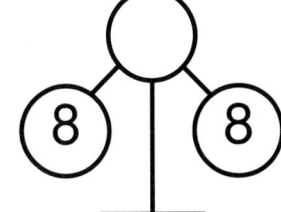 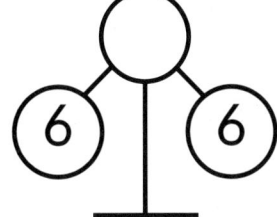

And can you work these out?

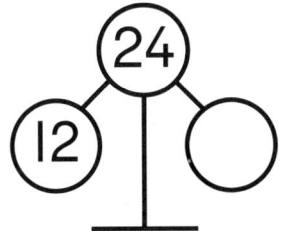  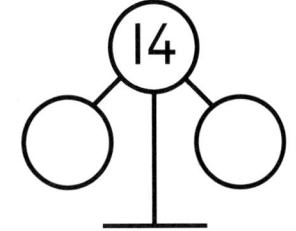 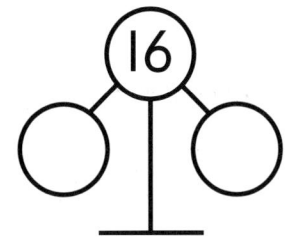

Buy two of each:

1 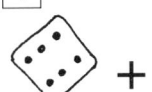 12p each Two cost 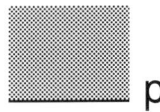 p

2  40p each Two cost 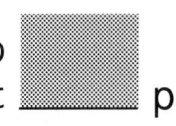 p

Dice doubles

1
 + = 12

6 + 5 =

6 + 7 =

2
+ = 10

5 + 6 =

5 + 4 =

3
+ = 8

4 + 5 =

4 + 3 =

8	Before you start IP 2	● Knows doubles of numbers to 10. Notes/date:

9

Name . . . . **Class** . . . .

Double it

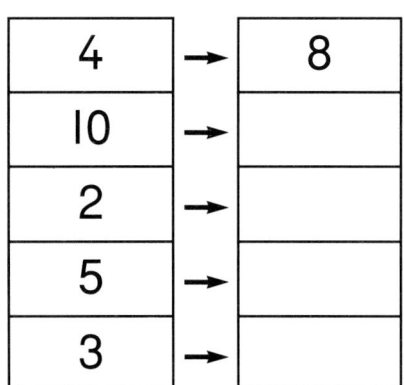

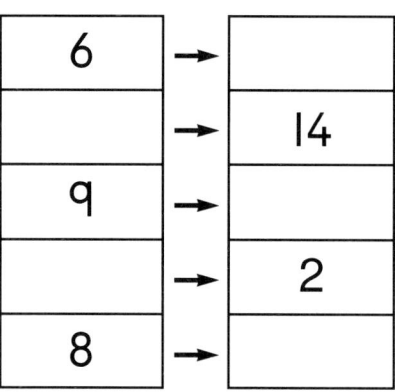

Doubles power

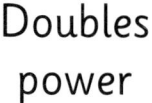
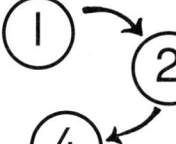

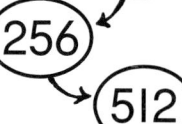

Use doubles to do these:

$(5 + 5) + 1 =$

$(3 + 3) - 2 =$

$(6 + 6) - 2 =$

$6 + (3 + 3) =$

$2 + (3 + 3) =$

$(8 + 8) - 1 =$

$(7 + 7) + 1 =$

$5 + (10 + 10) =$

Halving undoes doubling

5 and 5 = 10   Double 5 is 10   5 is half of 10

What is half of:

20   12    18    14

● Knows doubles for all numbers to 10.
Notes/date:

Before you start
IP 2

q

How many sides have:

Two triangles? $3 + 3 = 6$ sides

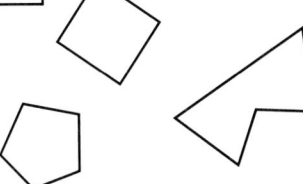

Two hexagons?    + ▨ = ▨ sides

Two squares?   ▨ + ▨ = ▨ sides

Two pentagons?   ▨ + ▨ = ▨ sides

Can you answer Nevis's questions?

What is double 9?

Two sevens are

$4 + 4 =$

Double 6?

Twice eight makes

Twice five is

 If you can do those, try my halving questions.

1 Twelve legs, how many birds?

2 Fourteen eyes, how many cats?

3 What's half of 24?

4 Sixteen beetles, half went away, how many left?

5 Twenty-two hands, how many children?

6 Half of 18 is

| 10 | Before you start IP 2 | ● Knows doubles of numbers to 12 and equivalent halves. Notes/date: |

What's half of 28 Netty?

Half of 20 is 10.
Half of 8 is 4.
So half of 28 is 10 + 4
and that is 14.

Try these:

half

14			21				
28	26	32	42	44	66	38	102

Double these numbers:

double

23	24	43	53	16	21	35	13	110
46								

Buy two of each:

1  90p   Two cost £ _____

2  65p   Two cost £ _____

3  72p   Two cost £ _____

4  £1·50   Two cost £ _____

Make up more:

5

6

- Can double numbers to 100 and above, and find equivalent halves.
- Can double money.
Notes/date:

Before you start
IP 2

11

Name . . . . **Class** . . . .

Which tens numbers are these between?

83 is between 80 and 90

16 is between 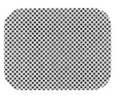 and

32 is between  and

28 is between [20] and [30] but nearer to [30]

19 is between [10] and  but nearer to

11 is between and [20] but nearer to

What is the <u>closest</u> tens number to:

99 [100]   8   51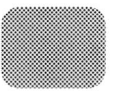

89    91    42

Round these numbers to the nearest ten:

7 [10]   18 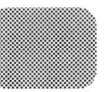   21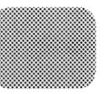

72    13    16

| 120 |
| 110 |
| 100 |
| 90 |
| 80 |
| 70 |
| 60 |
| 50 |
| 40 |
| 30 |
| 20 |
| 10 |
| 0 |

| 12 | Before you start IP 3 | ● Can round a two-digit number to the nearest 10. Notes/date: |

Pairs that make **20**

These are happy numbers because they add up to **20**

Make these numbers happy. They must add up to **20**

10 + ▨ = 20   9 + ▨ = 20   11 + ▨ = 20

5 + ▨ = 20   6 + ▨ = 20   4 + ▨ = 20

15 + ▨ = 20   ▨ + 4 = 20

▨ + 5 = 20   ▨ + 6 = 20

▨ + 7 = 20   ▨ + 8 = 20

▨ + 9 = 20   ▨ + 10 = 20

What will make these numbers happy?

8 + ▨ = 20   ▨ + 11 = 20

▨ + 18 = 20   7 + ▨ = 20

● Knows pairs of numbers with a total of 20.
Notes/date:

Before you start
IP 4

13

Name . . . . Class . . . .

What position was:

Norgay	Alice	Tracy	Jamell	Ben	Wendy	Ela	Dan	Ray
1st	2nd	3rd	4th	5th	6th	7th	8th	9th
First	Second	Third	Fourth	Fifth	Sixth			

What position was:

 Tracy? `3rd` _____  Ben? ☐ _____    Ela? ☐ _____  Norgay? ☐ _____

Position sentences:

Alice was second. There was 1 runner in front of her.

Dan was eighth. There were 7 runners in front of him.

Am I last again?

Chloe	Liz	Stan	Kit	Gail	Paul	Mick	Nevis
97th	98th	☐	☐	☐	☐	☐	104th

Write some position sentences:

_____

_____

_____

| 14 | Before you start IP 5 | ● Can compare/order numbers to at least 100. Notes/date: |

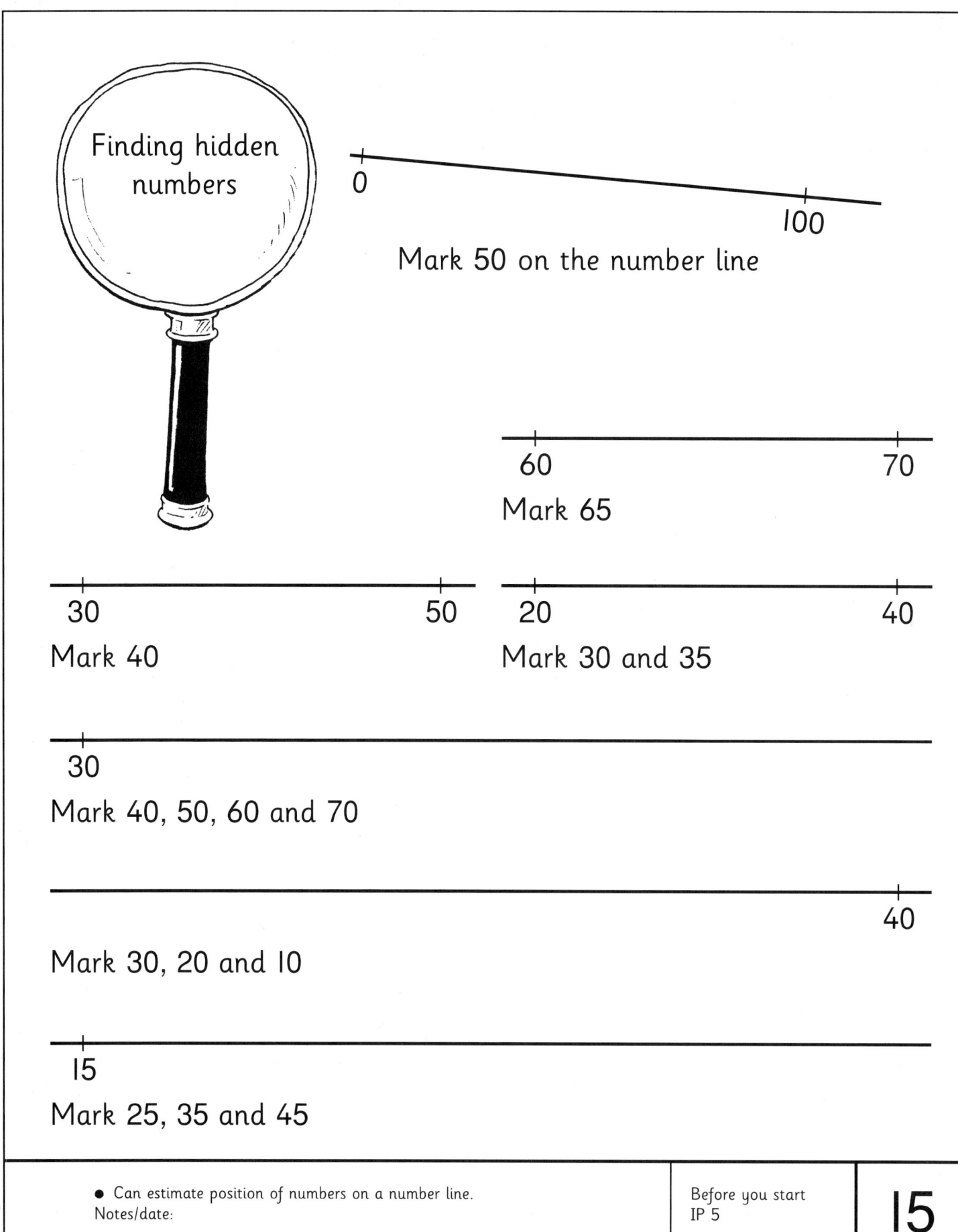

Finding hidden numbers

0   100

Mark 50 on the number line

60   70

Mark 65

30   50

Mark 40

20   40

Mark 30 and 35

30

Mark 40, 50, 60 and 70

40

Mark 30, 20 and 10

15

Mark 25, 35 and 45

● Can estimate position of numbers on a number line.
Notes/date:

Before you start
IP 5

15

Name . . . .   Class . . . .

You need
a partner
2 dice 1–6
a counter marked
9 on one side and
11 on the other side
a game track each

Come on!

Rules
- Take turns to throw the dice and counter.
- Add your three scores.
- Put a counter on the total.
- The winner is the first player to make a path across the river. The path can go up and down and sideways between circles that touch.

16	19
17	21	14
15	19	23
20	17	13
15	18

Name . . . . Class . . . .

Fill in the numbers on the numberlines:

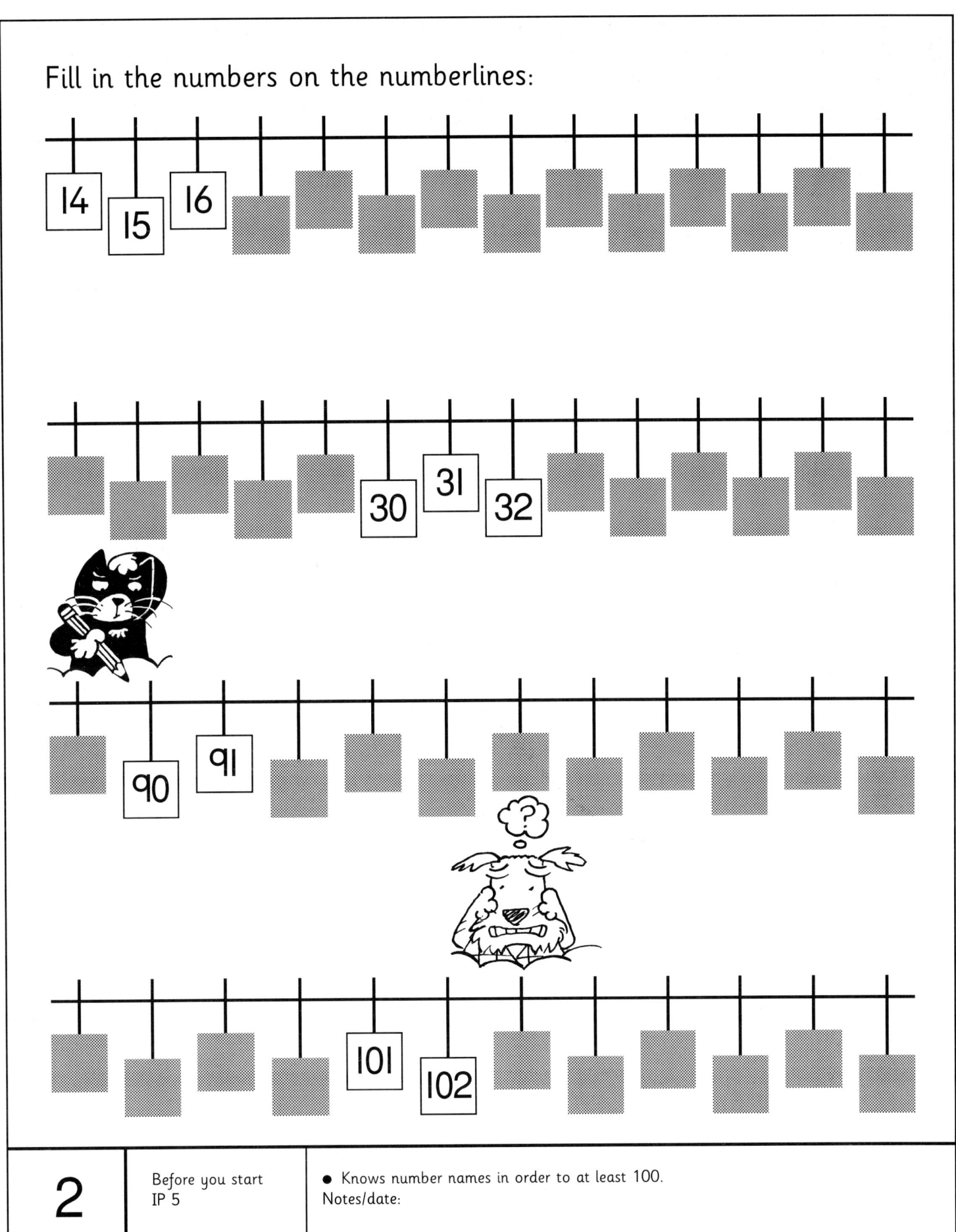

14 15 16

30 31 32

90 91

101 102

2   Before you start
IP 5

● Knows number names in order to at least 100.
Notes/date:

Make some jumps of 10.

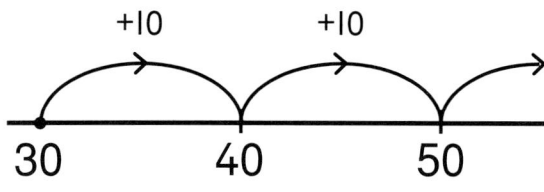

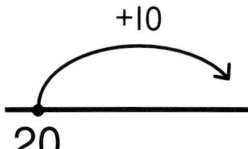

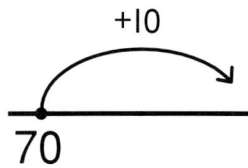

Now jump back in tens.

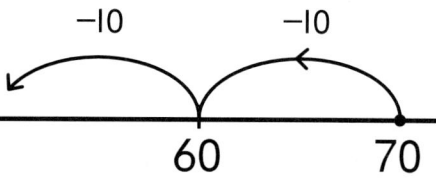

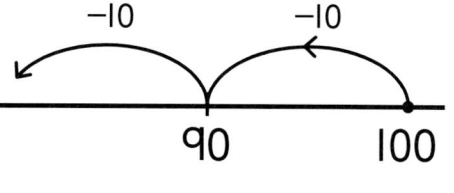

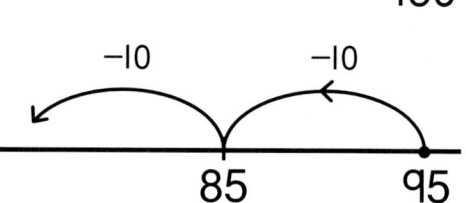

● Can count on/back in tens.
Notes/date:

Before you start
IP 6

3

Name . . . . Class . . . .

Make more jumps of 10.

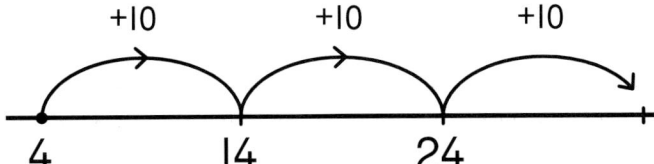

4   14   24

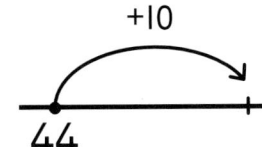

44

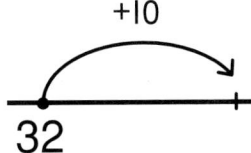

32

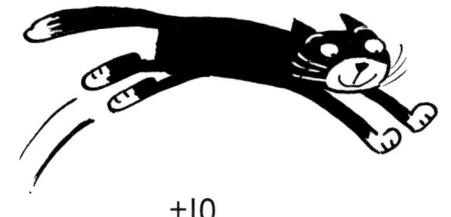

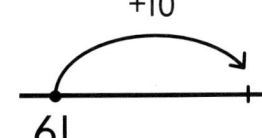

61

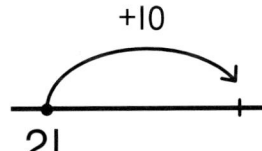

21

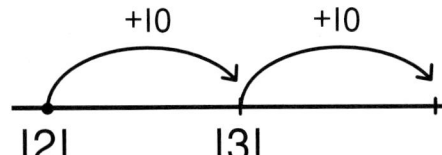

121   131

| 4 | Before you start
IP 6 | ● Can count on/back in tens from any number.
Notes/date: |

Name . . . . Class . . . .

Use jumping to add.

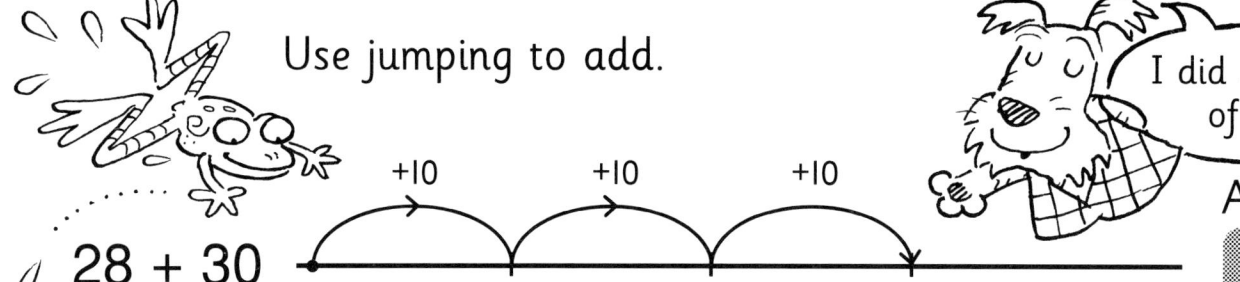

I did 3 jumps of ten.

Answer:

28 + 30   58

22 + 30 _____

15 + 40 _____

39 + 0 _____

Start at	Jump forward	Finish on
43	10	
54	20	
16	30	
12	40	
76	30	

28 + 10 = ____   48 + 10 = ____   68 + 10 = ____

38 + 20 = ____   58 + 10 = ____   78 + 10 = ____

● Can count on in tens from any number.
Notes/date:

Before you start
IP6

5

Jump back to subtract.

57 − 20

-10 -10

37   47   57

Answer:

 37

Start here!

65 − 40

65

77 − 40

77

89 − 30

44 − 30

Start at	Jump back	Finish on
83 − 10	=	
75 − 30	=	
88 − 20	=	
62 − 20	=	
91 − 40	=	
29 − 20	=	
100 − 50	=	
120 − 30	=	

6   Before you start   IP 6   ● Can count back in tens from any number.
Notes/date:

Split the numbers into three.

$7 + 2 + 1 = 10$

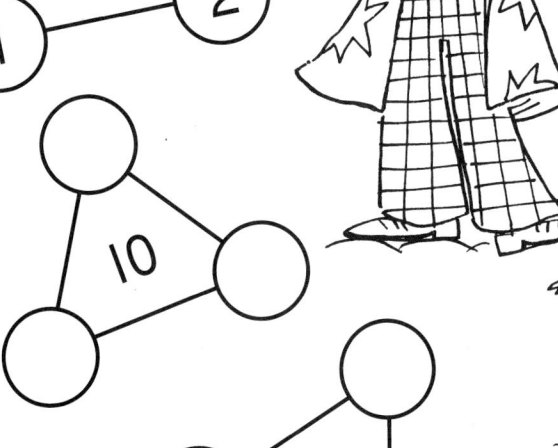

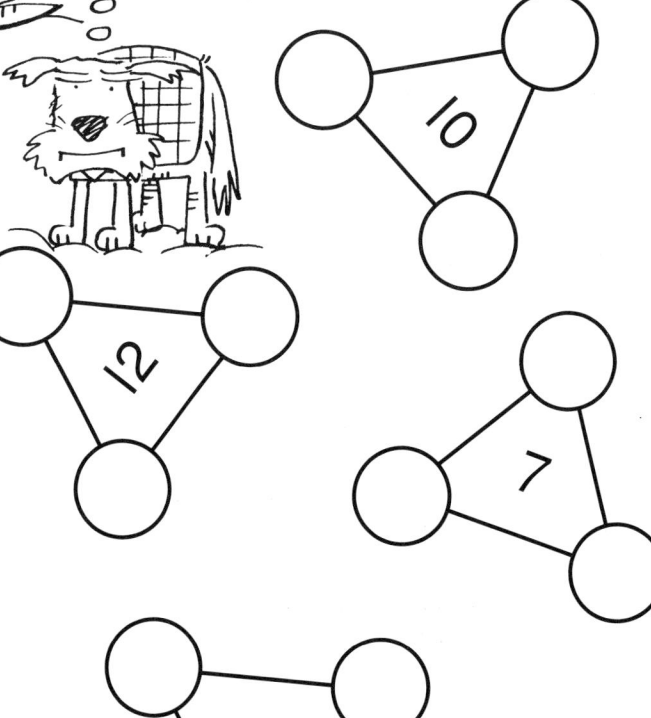

Make up more.

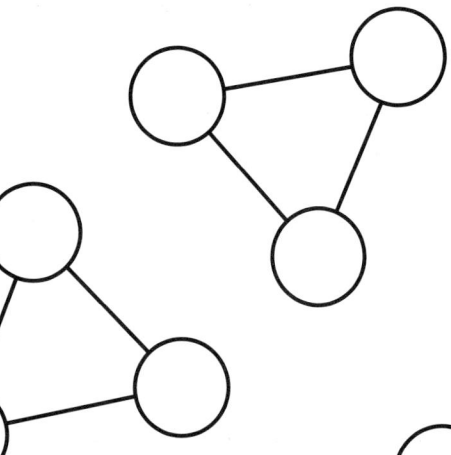

● Can partition and recombine numbers.
Notes/date:

Before you start
IP 7

7

Name . . . . **Class** . . . .

Go from 7 to 15.
You must use 2 jumps.

 Like this: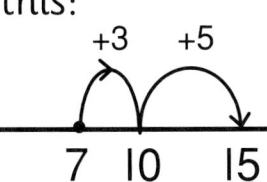

+3   +5

7   10   15

or this: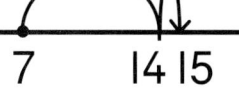

+7   +1

7   14 15

NOTICE
Only 2 jumps
allowed!

Make 2 jumps
from 5 to 14.

5

Make 2 jumps
from . . .   to . . .

3 → 11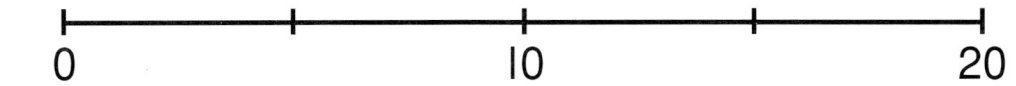
0   10   20

6 → 13
0   10   20

4 → 20
0   10   20

11 → 19
0   10   20

6 → 18
0   10   20

8 → 17
0   10   20

7 → 15
0   10   20

| **8** | Before you start IP 7 | ● Can partition numbers to aid mental addition. Notes/date: |

Name . . . .   Class . . . .

Show the sum on the number line:

Answer:

16 + 7 —————————————

12 + 9 —————————————

15 + 6 —————————————

14 + 8 —————————————

Show these subtractions:

————————————————→ 22 − 8
             22

————————————————→ 21 − 6

———————————————— 21 − 9

———————————————— 23 − 7

———————————————— 22 − 4

● Can partition numbers to aid mental addition/subtraction.
Notes/date:

Before you start
IP7

q

Dress costs £ ☐   Shoes costs £ ☐   Altogether they cost £ ☐

Number sentence: ☐

☐ marbles   He gives away ☐   ☐ left

Number sentence: ☐

☐ on the bus   ☐ get on   ☐ get off   ☐ on the bus now

Number sentence: ☐ + ☐ – ☐ = ☐

10	Before you start IP 8	● Understands when to use addition and subtraction and the related vocabulary. Notes/date:

Name . . . . **Class** . . . .

How many candles will they have?

 McMagic
 Netty
 Kit
Nevis
68 years old   12 years old   14 years old   6 years old

Number of cards

McMagic 72
Nevis 6
Netty 30
Kit 3

Nevis and Kit together?  +  =  candles

Kit and Netty together?  + ☐ =

McMagic and Nevis?  + ☐ = ☐

Nevis, Kit and Netty?  +  +  =

Kit is  years older than Nevis | 14 | − | 6 | =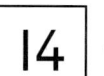

Write sentences about the cards they received.

_____

_____

_____

_____

_____

● Can solve numerical problems in context.
Notes/date:

Before you start
IP 8

11

Name . . . . **Class** . . . .

What story does this line tell?

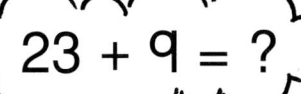

23 + 9 = ?

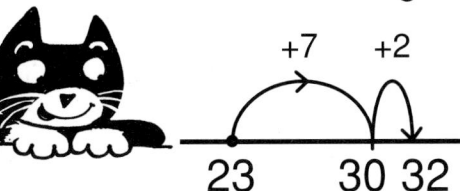

+7   +2

23   30 32

Kit started on 23, added 7 to make 30, and then 2 to make 32!

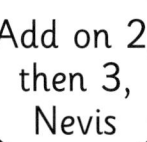

Add on 2, then 3, Nevis

Help Nevis with these

18 + 5 = ?

18

17 + 4 = ?

17

I'd add 3 first!

18 + 6 = ?

16 + 5 = ?

Try not to count in ones, Nevis!

What should I do then?

Use your number pairs to go to the nearest ten!

| 12 | Before you start
IP 9 | ● Can look for number pairs that make ten to aid addition.
Notes/date: |

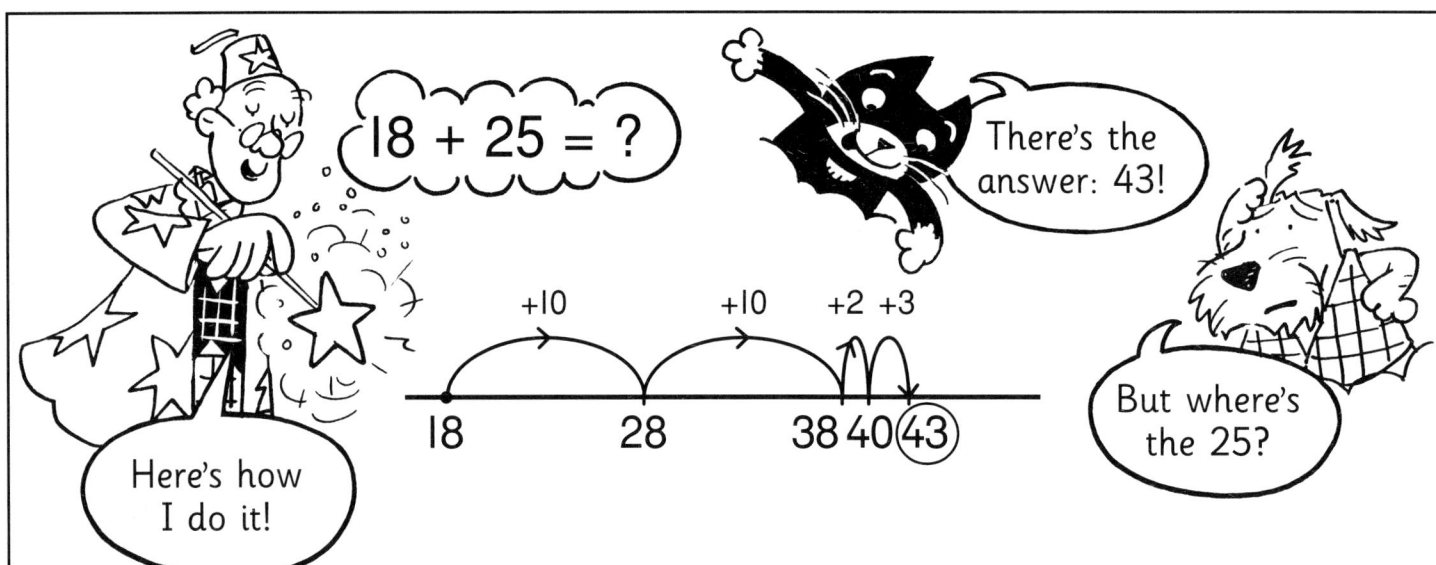

Write the number line stories for:

20 + 16 = ?

20

22 + 27 = ?

54 + 9 = ?

27 + 24 = ?

35 + 18 = ?

● Can partition addition into tens and units, then recombine.
Notes/date:

Before you start
IP 9

13

Name . . . . **Class** . . . .

Use the empty number line to subtract.

Jump back 13 from 54:

$54 - 13$ is ▨

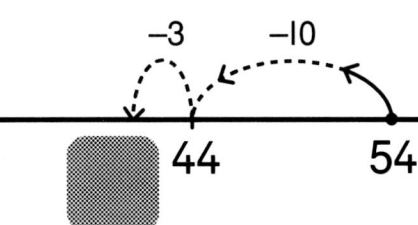

44   54

Jump back 12 from 38:

$38 - 12 =$ ▨

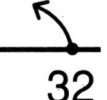

38

$32 - 21 =$ ▨

32

Make up some more!
Show me what you can do!

▨ − ▨ = ▨   _____

▨ − ▨ = ▨   _____

▨ − ▨ = ▨   _____

▨ − ▨ = ▨   _____

| 14 | Before you start
IP 9 | ● Can subtract by subtracting tens and then units.
Notes/date: |

Name . . . . Class . . . .

Draw on the number line:

4 sets of two:

$4 \times 2 =$ ▢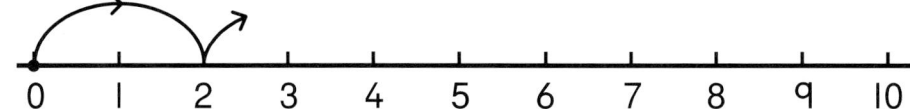

0 1 2 3 4 5 6 7 8 9 10

Two groups of 5:

$2 \times 5 =$ ▢

0 1 2 3 4 5 6 7 8 9 10 11 12

Six groups of 3:

$6 \times 3 =$ ▢

0 1 2 3 4 5 6 7 8 9 10 11 12 13 14 15 16 17 18 19 20

Write the number sentence

3 lots of 3 = ▢

$3 \times 3 =$ ▢

3 lots of 2 = ▢

▢ $\times 2 =$ ▢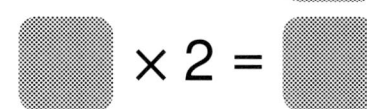

$2 \times$ ▢ $=$ ▢

▢ $\times 3 =$ ▢

● Can understand the operation of multiplication and related vocabulary.
Notes/date:

Before you start
IP 10

15

Mrs Lloyd has 12 sheep.

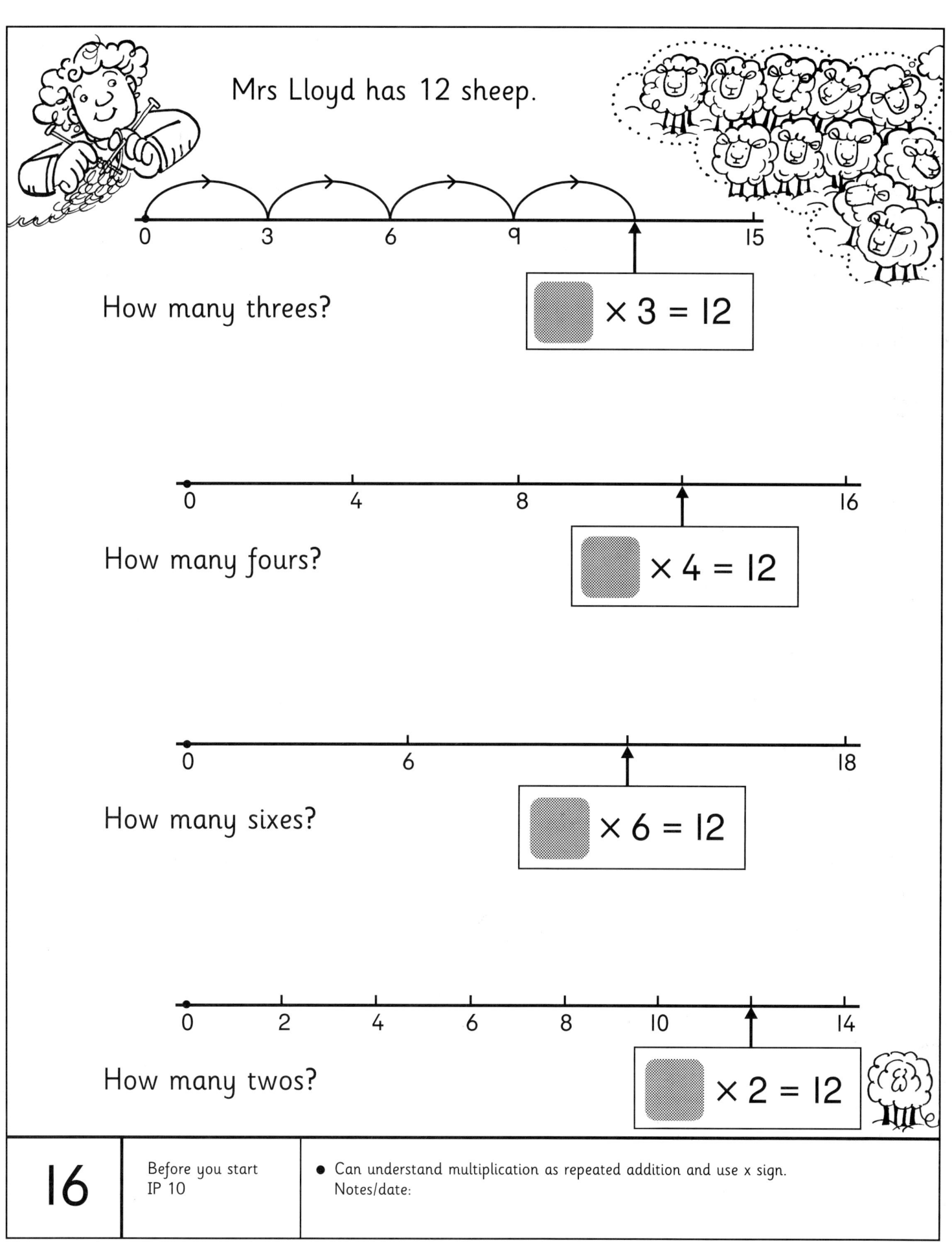

How many threes?

$\boxed{} \times 3 = 12$

How many fours?

$\boxed{} \times 4 = 12$

How many sixes?

$\boxed{} \times 6 = 12$

How many twos?

$\boxed{} \times 2 = 12$

16	Before you start IP 10	● Can understand multiplication as repeated addition and use x sign. Notes/date:

Kit jumps in fours.

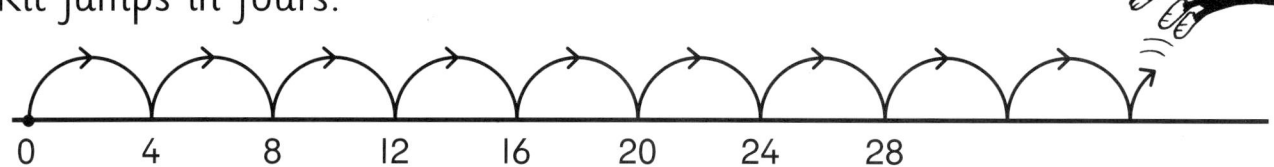

0   4   8   12   16   20   24   28

How many jumps make 20?    jumps

 × 4 = 20

 × 4 = 8    × 4 = 4    × 4 = 36

 × 4 = 24    × 4 = 16    × 4 = 0

 × 4 = 12    × 4 = 3    × 4 = 28

More multiplying

Psst! Use doubling for these!

2 cost 2 × 3p =  p

4 cost 4 × 3p =  p

8 cost 8 × 3p =  p

3 cost 3 × £2 = £

6 cost 6 × £2 = £

12 cost 12 × £2 = £

5 cost 5 × £4 = £

10 cost 10 × £4 = £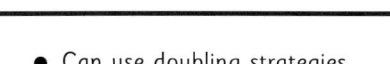

20 cost 20 × £4 = £

4 cost 4 × £5 = £

8 cost 8 × £5 = £

16 cost 16 × £5 = £

● Can use doubling strategies.
Notes/date:

Before you start
IP 10

17

Name . . . . Class . . . .

Help Nevis by filling in the house numbers.

66   68

Brook Street

59 Brook St.   63   67

Put a **circle** around all the **odd** numbers

7   52   146   232   63   88
        15   14
12   77   46   36   159   99
        35   24
21   85   436   51
    138   79   47   49

Put a cross through all the **even** numbers.

How many circles? [ ]   Is that odd or even? [ ]

Sort the numbers into the boxes.

odd numbers over 100

odd numbers smaller than 100

even numbers over 100

even numbers less than 70

Name . . . .   **Class** . . . .

Take 4 cubes and make a square!

Can you make a square with 6 cubes?

4 <u>is</u> a square number.
Make a square with a different number of cubes.

Is six a square number?

Use cubes to find the square numbers.

1	2	3	4	5	6	7	8	9	10
✓					✗				

How many beetles in each square?

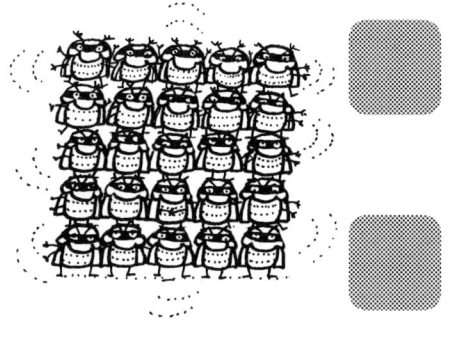

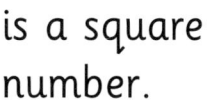 beetles

 is a square number.

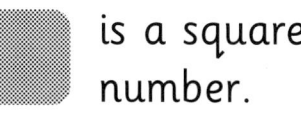 beetles

 is a square number.

● Can begin to recognise square numbers less than 100.
Notes/date:

Before you start
IP 11

19

Name . . . . **Class** . . . .

Draw a circle round the multiples of 2.

Draw a cross on the multiples of 10.

0	1	2	3	4	5	6	7	8	9
10	11	12	13	14	15	16	17	18	19
20	21	22	23	24	25	26	27	28	29
30	31	32	33	34	35	36	37	38	39
40	41	42	43	44	45	46	47	48	49
50	51	52	53	54	55	56	57	58	59
60	61	62	63	64	65	66	67	68	69
70	71	72	73	74	75	76	77	78	79
80	81	82	83	84	85	86	87	88	89
90	91	92	93	94	95	96	97	98	99

How many tens?

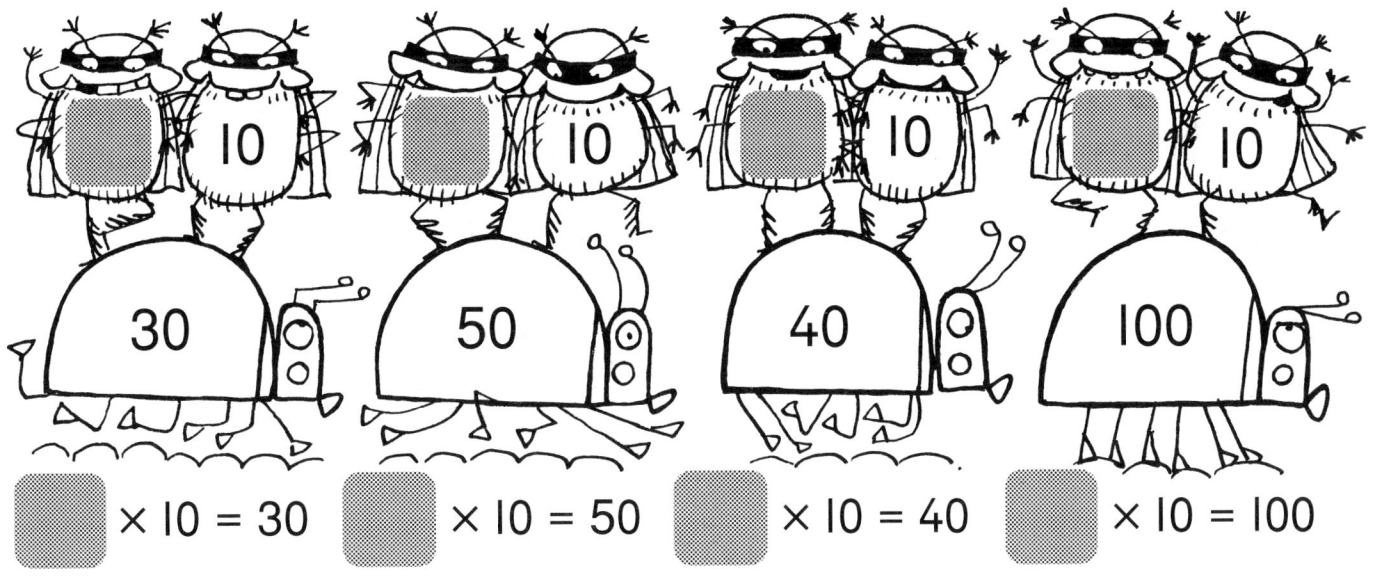

☐ × 10 = 30   ☐ × 10 = 50   ☐ × 10 = 40   ☐ × 10 = 100

20	Before you start IP 12	● Can recognise multiples of 2 and 10. Notes/date:

Name . . . .   Class . . . .

Colour the multiples of 3 like this:

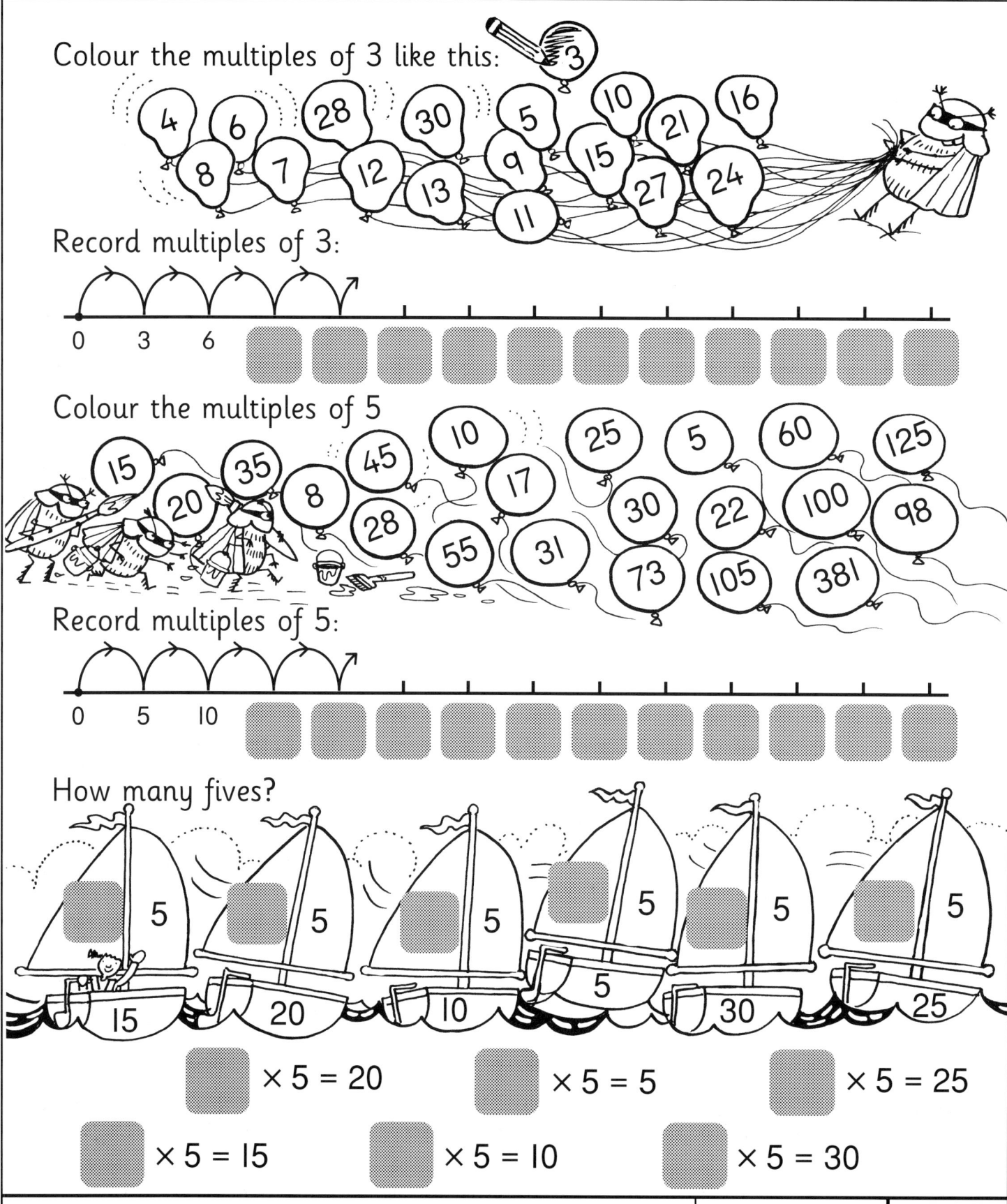

Record multiples of 3:

0   3   6

Colour the multiples of 5

Record multiples of 5:

0   5   10

How many fives?

15   20   10   30   25

☐ × 5 = 20   ☐ × 5 = 5   ☐ × 5 = 25

☐ × 5 = 15   ☐ × 5 = 10   ☐ × 5 = 30

● Can recognise multiples of 3 and 5.
Notes/date:

Before you start
IP 12

21

Name . . . . Class . . . .

Using multiples of 10 and 100

Look hard and find the pattern!

$4 + 5 = 9 \rightarrow 40 + 50 = 90 \qquad 400 + 500 = 900$

$2 + 3 = 5 \qquad 20 + 30 = 50 \qquad 200 + 300 = 500$

$6 + 3 = \boxed{} \qquad 60 + 30 = \boxed{} \qquad 600 + 300 = \boxed{}$

 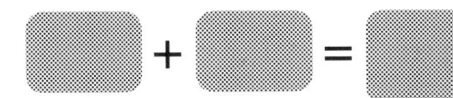

$8 - 6 = 2 \qquad 80 - 60 = \boxed{} \qquad 800 - 600 = \boxed{}$

$7 - 5 = \boxed{} \qquad 70 - 50 = \boxed{} \qquad 700 - 500 = \boxed{}$

$9 - 3 =$

  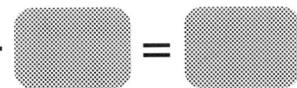

22	Before you start IP 13	● Can recognise and use patterns of multiples in calculations.. Notes/date:

Jump in tens to answer these:

$27 + 40 =$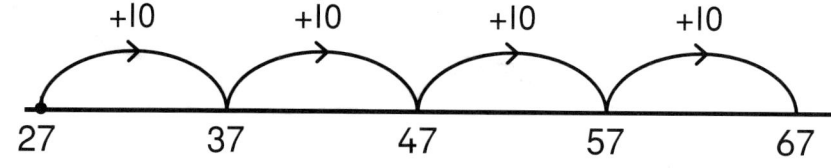

$53 + 30 =$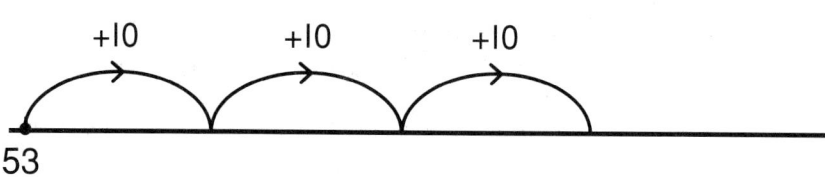

$103 + 70 =$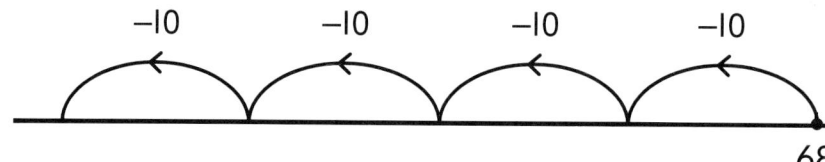

$68 - 40 =$

$88 - 50 =$

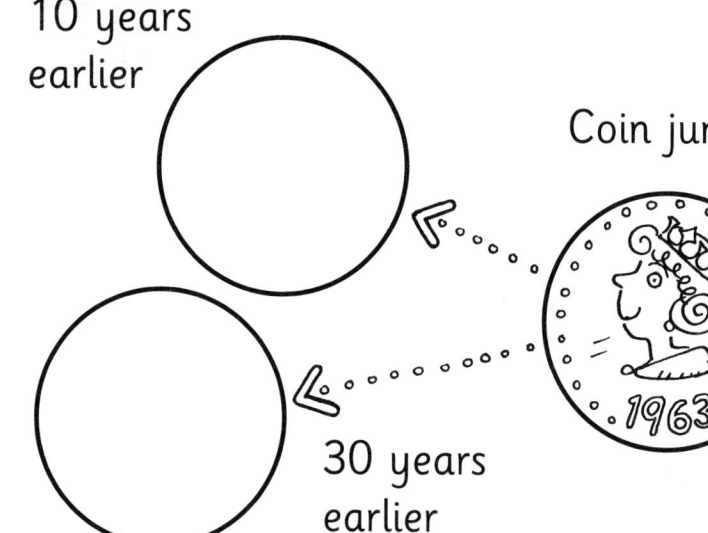

Coin jumps

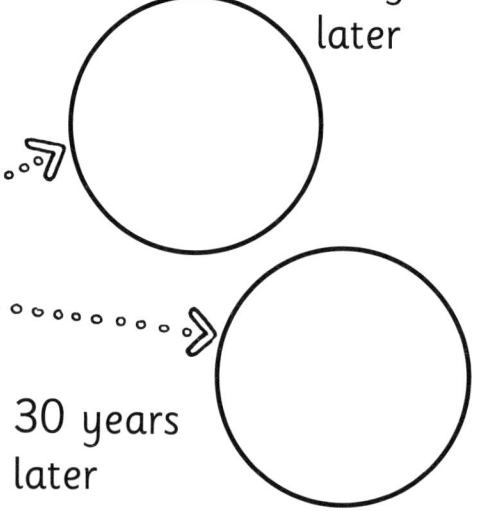

10 years earlier

20 years later

30 years earlier

30 years later

● Can add and subtract multiples of 10 to/from any number.
Notes/date:

Before you start
IP 13

23

Name . . . . **Class** . . . .

Division can mean putting things into groups

 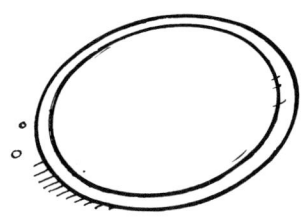

Put out the biscuits, 4 to a plate.
How many plates? Draw them.

How many 4s in 20?

20 ÷ 4 =

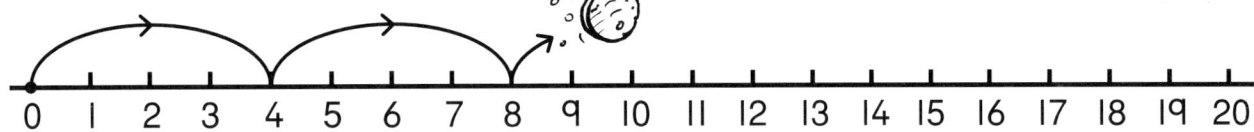

12 ÷ 4 =   24 ÷ 4 =

5 cups on a tray.
How many trays for 20 cups?

20 ÷ 5 =

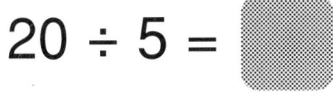

15 ÷ 5 =   30 ÷ 5 =

24	Before you start IP 14	● Can understand division as equal groupings. Notes/date:

Division can mean <u>sharing</u>.
Share the 24 apples fairly between 4.

24 apples

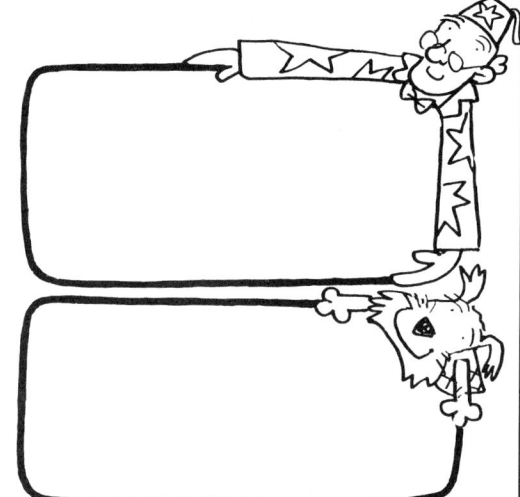

How many each? ▢

$24 \div 4 = \boxed{}$

$4 \times \boxed{} = 24$

Share the 18 sandwiches
fairly between 3.

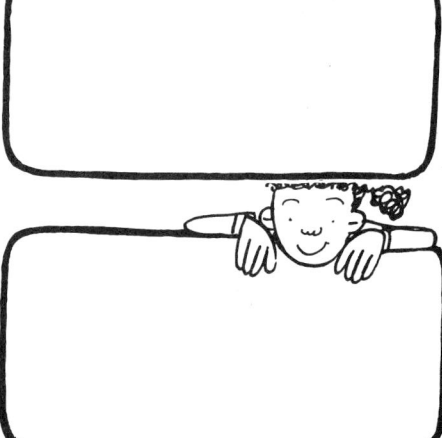

How many each? ▢

$18 \div 3 = \boxed{}$

$3 \times \boxed{} = 18$

● Can understand division as sharing equally.
Notes/date:

Before you start
IP 14

25

Name . . . . Class . . . .

1 How many 2s make 20?

$20 \div 2 = $ ⬚ and ⬚ $\times 2 = 20$

2 How many 3s make 21?

⬚ $\times 3 = 21$ and $21 \div 3 = $ ⬚

Remainders

7 seeds
Each Bandit beetle gets 2

2 each, and 1 left over

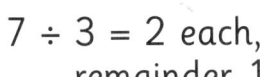

$7 \div 3 = 2$ each, remainder 1 .

Find the remainder

1 10 seeds
3 beetles

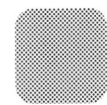 each, remainder

2 19 seeds, 2 beetles

 each, remainder

You can use cubes for seeds!

| 26 | Before you start IP 14 | ● Can begin to understand and read 'remainder' as what is left after sharing out. Notes/date: |

Do the opposite to check if it's true!  $9 + 2 = 11$

True or not true?  $11 - 9 = 2?$

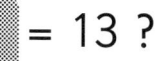 True

Check subtraction by <u>adding</u> 9 and 2.

Use adding to check these subtractions:

$13 - 9 = \boxed{}$ True / not true? Adding check: Is $9 + \boxed{} = 13$?

$20 - 10 = \boxed{}$ Adding check: Is $\boxed{} + 10 = 20$?

$14 - 8 = \boxed{}$ Adding check: Is $8 + \boxed{} = 14$?

Check addition by adding in a different order:

$4 + 3 + 2 = \boxed{}$ Is $2 + 3 + 4 = \boxed{}$

Check these by adding in a different order:

$3 + 7 + 8 = \boxed{}$ Check:

$5 + 4 + 3 + 3 = \boxed{}$ Check:

$16 + 2 + 4 = \boxed{}$ Check:

$5 + 6 + 8 = \boxed{}$ Check:

● Can check the result of a calculation by using the inverse operation (+/−).
Notes/date:

Before you start
IP 15

27

These sentences are <u>not true</u>. Explain how you know.

$56 - 10 = 44$ <u>Not true because</u> . . .
No ✗

When you subtract a ten the units number will stay the same!

47 is in the two times table.
No ✗

<u>Not true because</u> . . .

136 is a multiple of 5.
No ✗

<u>Not true because</u> . . .

$43 + 20 = 66$
No ✗

<u>Not true because</u> . . .

$20 - 4 = 15$
No ✗

<u>Not true because</u> . . .

637 is even.
No ✗

<u>Not true because</u> . . .

$8 + 9 = 16$
No ✗

<u>Not true because</u> . . .

| 28 | Before you start IP 15 | • Can explain methods and reasoning about numbers. Notes/date: |

1  Altogether  pens

2  + =

3 How much? £

4 What number is double 9 ?

5

6

7

8

9 What number is half of 24 ?

● Can use the appropriate operation to solve problems.
Notes/date:

Before you start
IP 16

29

Name . . . . Class . . . .

Solve these puzzles – then share your secrets!

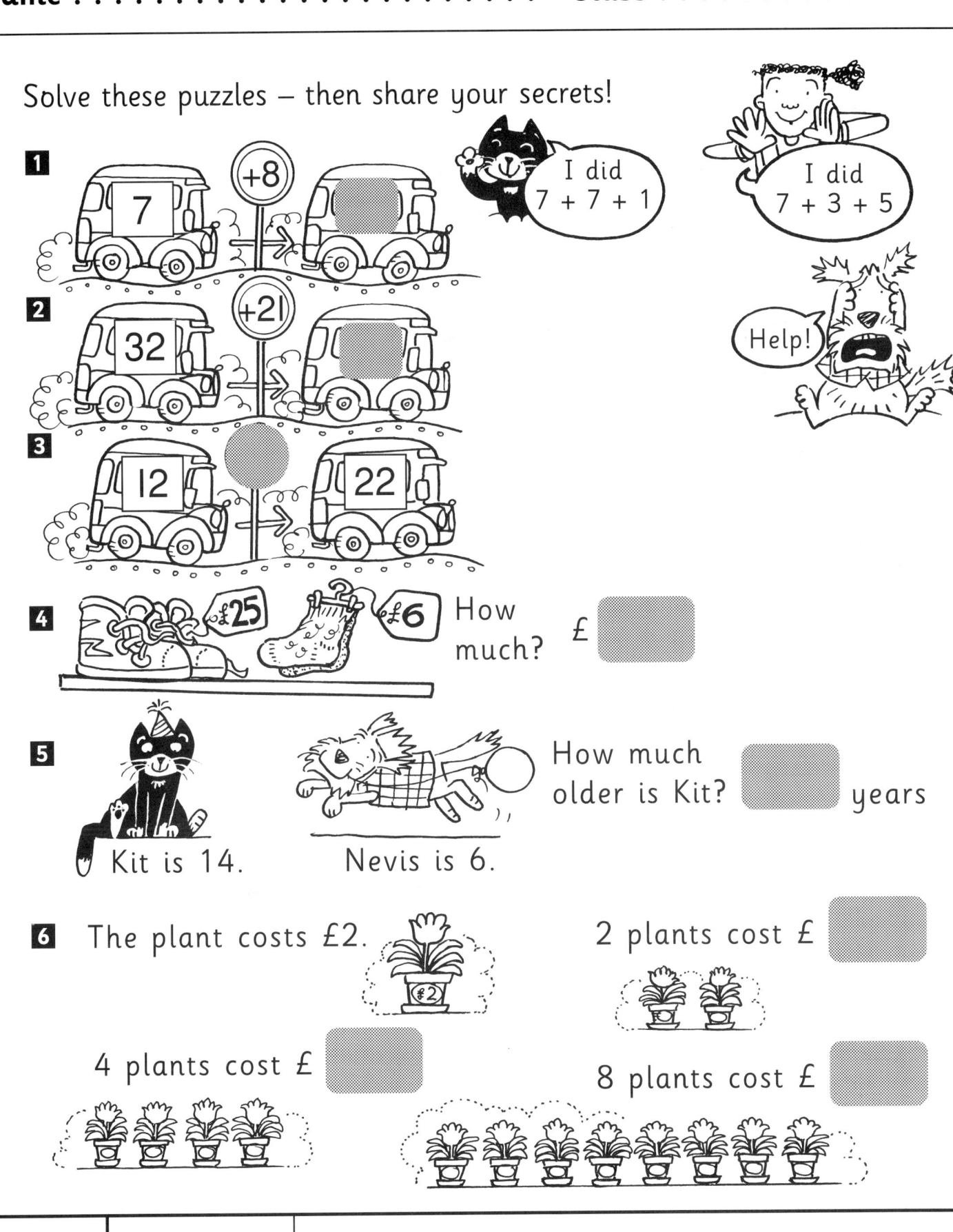

I did
7 + 7 + 1

I did
7 + 3 + 5

Help!

1 7 +8

2 32 +21

3 12 22

4 £25 £6 How much? £ ▢

5 How much older is Kit? ▢ years

Kit is 14. Nevis is 6.

6 The plant costs £2.

2 plants cost £ ▢

4 plants cost £ ▢

8 plants cost £ ▢

30	Before you start IP 16	● Can use the appropriate operation to solve problems and explain methods orally. Notes/date:

Name . . . . **Class** . . . .

Solve these stories – then share your secrets!

1 Nevis collected 16 bones. He buried 9. How many were left?

2 There are 19 children on the top deck and 11 on the lower deck of the bus. How many on the bus altogether?

3 Netty is 12 years old. McMagic is 68. How much older is McMagic?

4 Kit, Netty and Nevis have 4 cakes each. How many is that?

5 The eggbox holds 6 eggs. How many eggs in 4 boxes?

How many boxes do you need for 30 eggs?

6 I thought of a number. Then I halved it. The answer was 6. What was the number?

● Can solve 'story' problems related to real life and explain methods orally.
Notes/date:

Before you start
IP 16

31

Practice Book 3

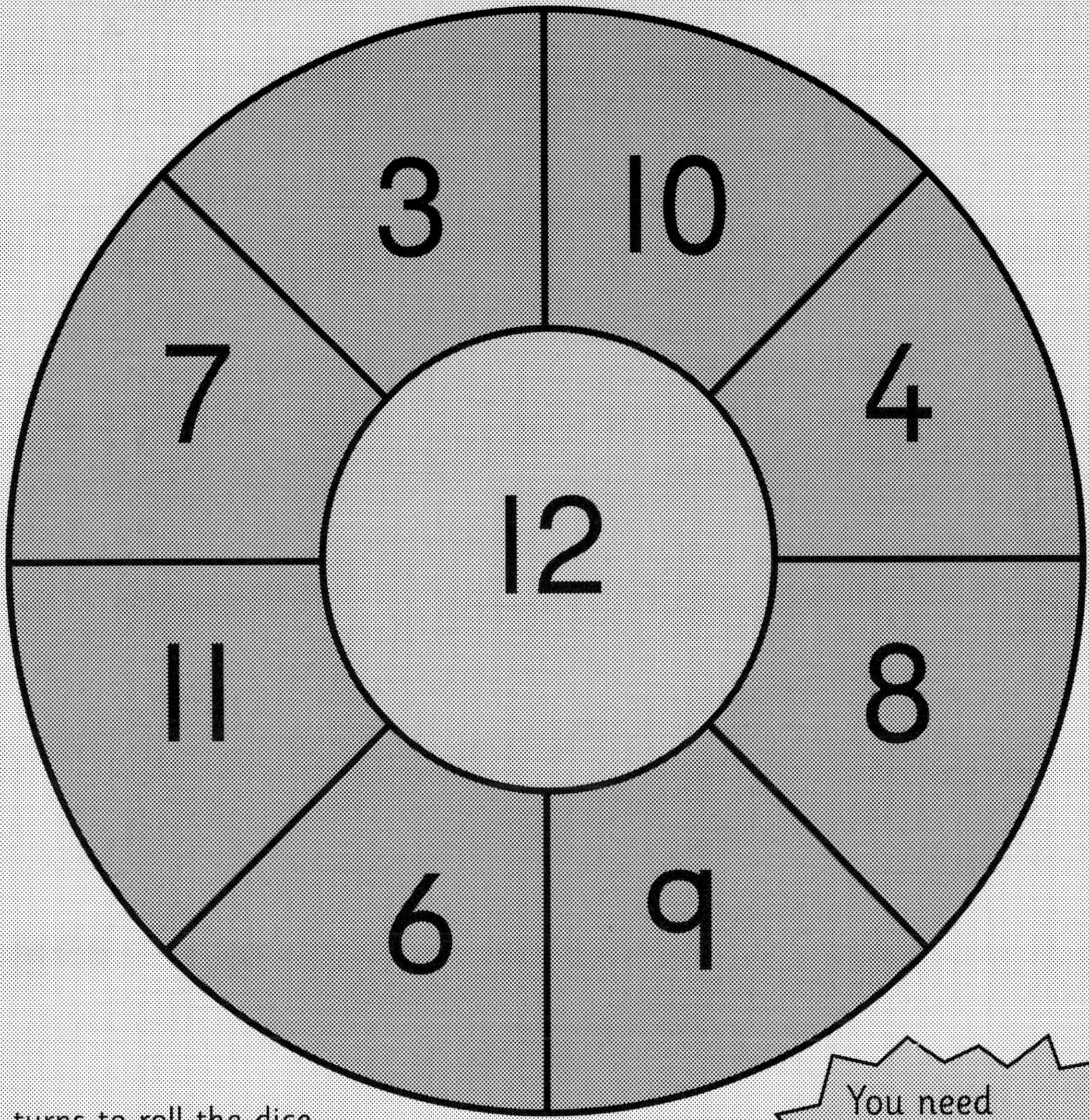
3
10
7
4
12
11
8
6
9

Rules

- Take turns to roll the dice.
- Add the two numbers.
- If your total is on the board and doesn't already have one of your counters, put a counter in the space.
- The winner is the first player to get one of their counters in each space.

You need
2 players
2 dice 1–6
9 counters each, a different colour for each player

Name . . . . **Class** . . . .

Rules
- Take turns to roll the dice.
- Add the three numbers.
- If your total is on the board and doesn't already have one of your counters, put a counter in the space.
- The winner is the first player to get one of their counters in each space.

You need
2 players
3 dice 1–6
9 counters each,
a different colour
for each player

Name . . . . Class . . . .

1 ☐   4 ☐

2 ☐   5 ☐

3 ☐   6 ☐

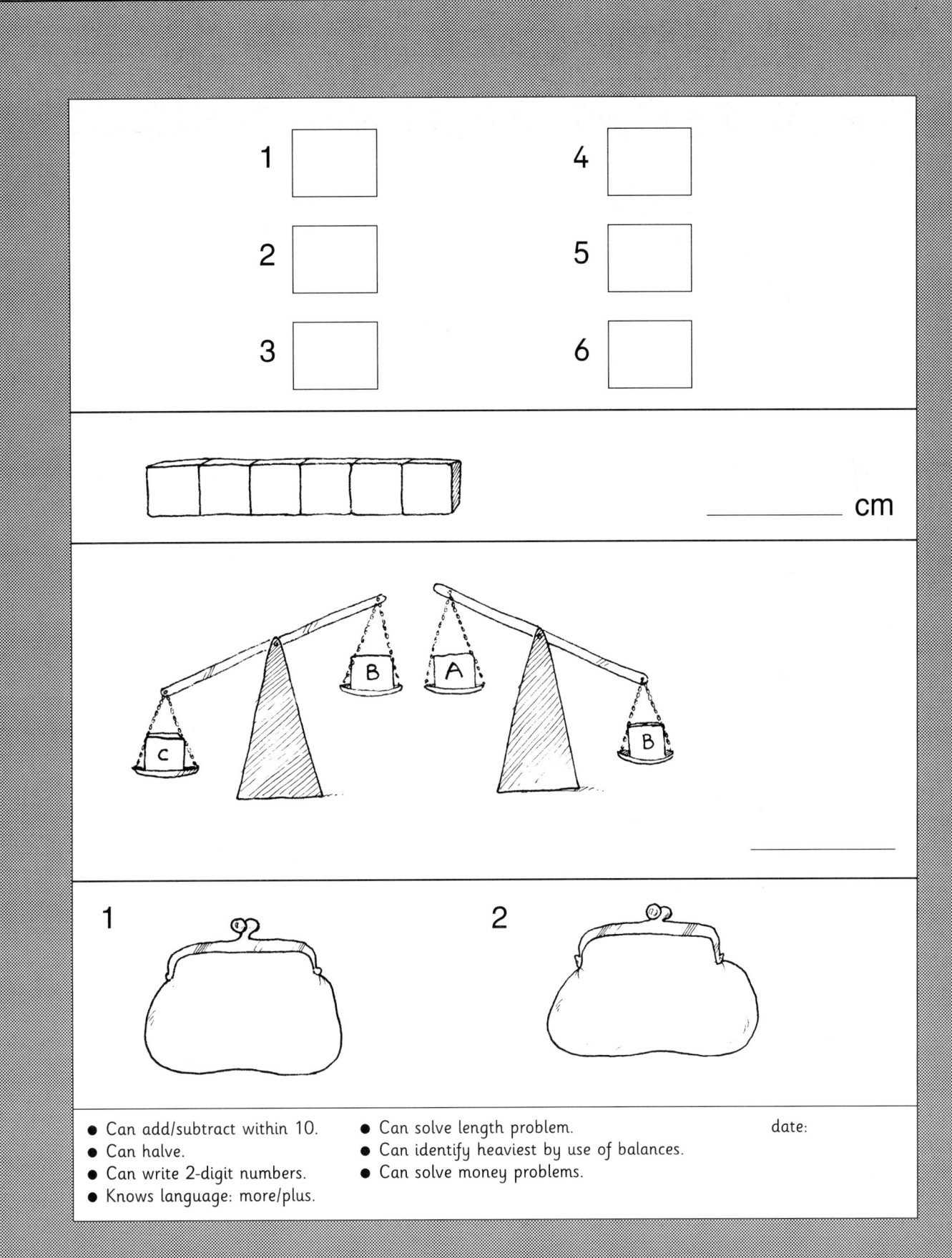

_____ cm

1   2

● Can add/subtract within 10.   ● Can solve length problem.   date:
● Can halve.   ● Can identify heaviest by use of balances.
● Can write 2-digit numbers.   ● Can solve money problems.
● Knows language: more/plus.

2

Name . . . .   Class . . . .

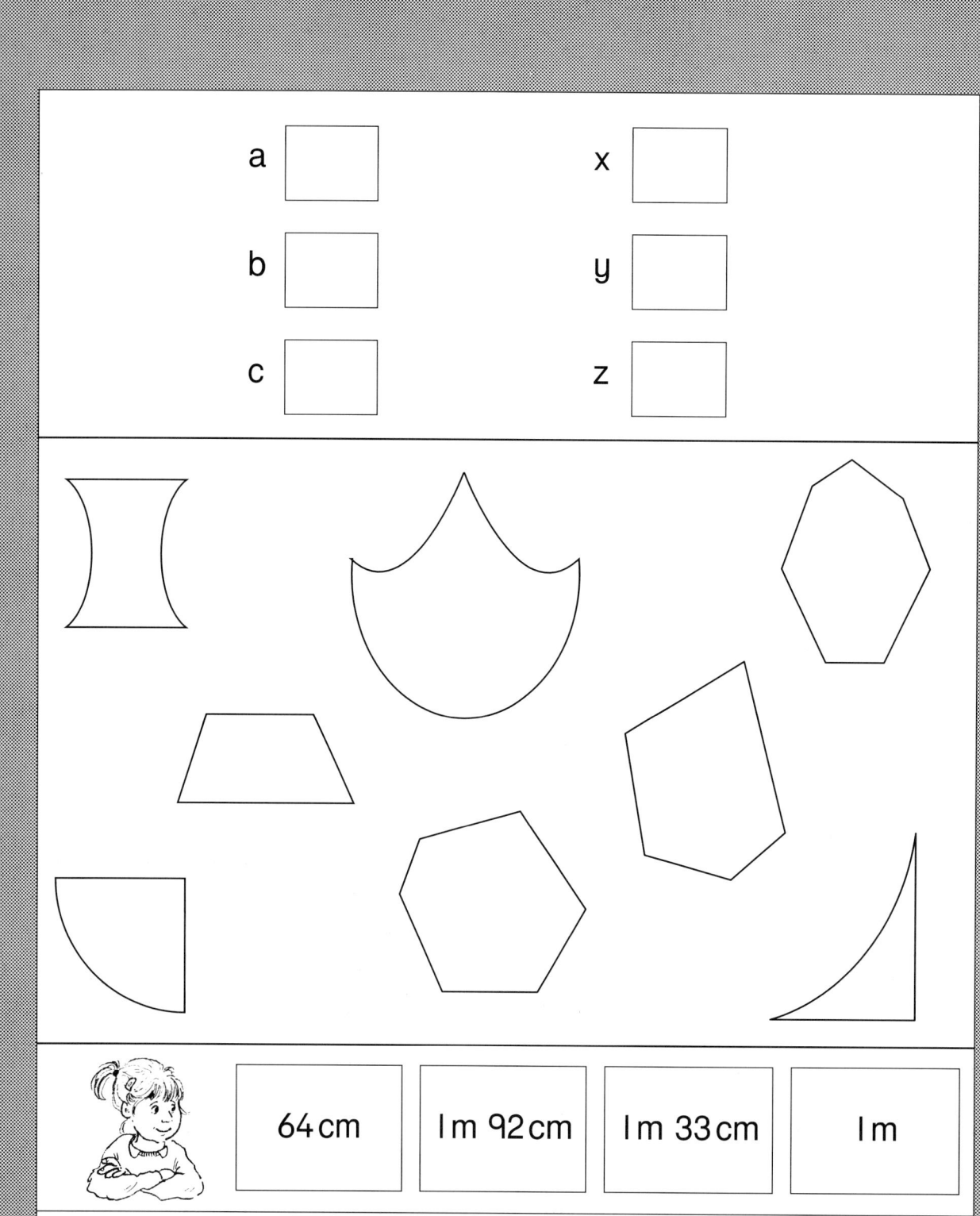

a ☐   x ☐

b ☐   y ☐

c ☐   z ☐

64 cm   1 m 92 cm   1 m 33 cm   1 m

- Can write 2-digit number.
- Can add/subtract within 20.
- Can double.
- Knows odd/even.

- Can identify shape by properties.
- Can identify a right angle.
- Can choose appropriate length of measure.

date:

3

Name . . . . **Class** . . . .

1 _____   6 _____

2 _____   7 _____

3 _____   8 _____

4 _____   9 _____

5 _____   10 _____

200 grams   _____ apples

- Can add/subtract/multiply within 20.
- Can sequence 2-digit numbers.
- Can continue a number pattern.
- Can double/halve.
- Understands 'less than'.
- Can add/subtract 10.
- Can solve gram/kilogram problem.
- Can select coins.

date:

4

53

Name . . . . Class . . . .

1 ☐   4 ☐

2 ☐   5 ☐

3 ☐   6 ☐

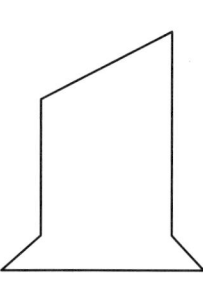   200 ml   _____

$$5 + 4 \qquad 5 - 4 \qquad 5 \times 4 \qquad 4 + 5$$

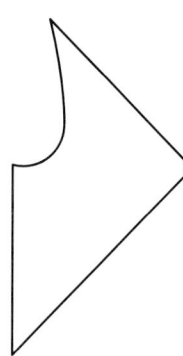 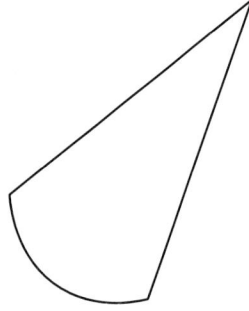 

● Can add/subtract/multiply.
● Can solve division problem.
● Can double.
● Understands and uses 'less than'.

● Can solve ml/litre problem.
● Can choose appropriate operation to solve a problem.
● Can identify reflective symmetry/right angle.

date:

5

Name . . . . Class . . . .

1 _____	6 _____
2 _____	7 _____
3 _____	8 _____
4 _____	9 _____
5 _____	10 _____

February   October   July   May   November   January   June

_____ cm

- Can add/subtract/multiply.
- Can solve money problem.
- Can write 3-digit numbers.
- Can continue a number pattern.
- Can find $\frac{1}{4}$ of a quantity.

- Can halve/double.
- Understands and uses language of number.
- Can sequence months.
- Can use multiplication to solve a problem.

date:

6

Name . . . . **Class** . . . .

Match 16 to
3 other boxes.

18 – 3

10 + 6

7 + 4 + 5

6 + 5 + 6

16

9 + 8

5 × 3

20 – 4

4 × 4

Complete this sequence of numbers.

5	7	9		13				19		

66 $\xrightarrow{\text{10 more}}$

75 $\xrightarrow{\text{10 less}}$

27 $\xrightarrow{\text{10 less}}$

84 $\xrightarrow{\text{10 more}}$

30 + 20 =

60 – 30 =

52 + 9 =

12 + 11 =

- AT2 level 2: recognises sequences of numbers.
- Can add/subtract within 20.
- Understands more/less.
- Uses mental methods to add/subtract 10.
- Can add/subtract 2-digit numbers.

date:

7

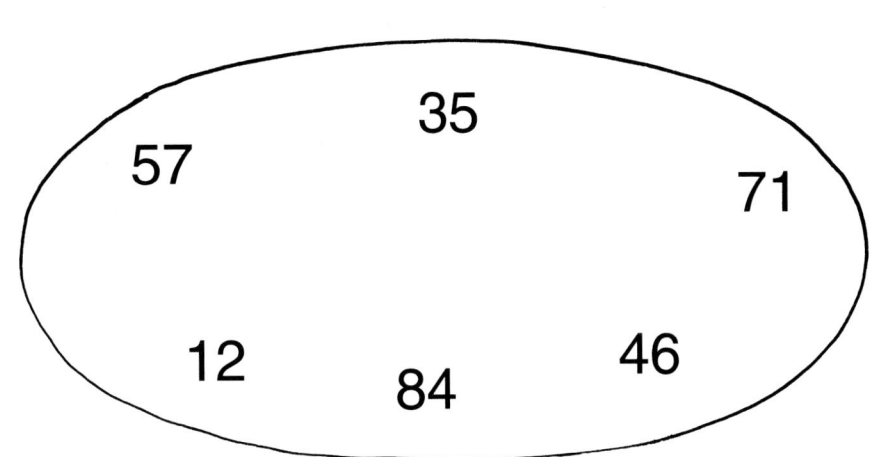

Look at these numbers.

● Write the numbers in order.

smallest

● Write the numbers that are more than 50.

_____

● Write the numbers that are less than 40.

_____

● Which number is the largest?

_____

● Put a ring round the even numbers.

8

● AT2 level 2: understands place value of and can order numbers to 100.
● AT2 level 2: recognises even numbers.
● AT1 PoS: can use the language of number.

date:

Name . . . . **Class** . . . . q . .

Put the numbers in the boxes to complete the sums.

$2 + 4 = \boxed{}$ $6 + 7 = \boxed{}$

$5 + 1 = \boxed{}$ $9 + 8 = \boxed{}$

$7 + 0 = \boxed{}$ $\boxed{} + \boxed{} = 20$

$\boxed{} + \boxed{} = 8$ $20 - 5 = \boxed{}$

$\boxed{} + \boxed{} = 10$ $12 - 4 = \boxed{}$

$5 - 1 = \boxed{}$ $\boxed{} - \boxed{} = 10$

$7 - 2 = \boxed{}$ $6 \times 2 = \boxed{}$

$9 - 4 = \boxed{}$ $3 \times 3 = \boxed{}$

$\boxed{} - \boxed{} = 1$ $\boxed{} \times \boxed{} = 20$

$\boxed{} - \boxed{} = 3$ $\boxed{} \times \boxed{} = 15$

- AT2 level 2: mental recall of + and − facts to 10.
- AT2 level 3: mental recall of + and − facts to 20.
- AT2 level 3: mental recall of 2, 5 and 10 × tables and others to 5 × 5.

date:

q

Name . . . . **Class** . . . .

How many fish? _____

If 3 swam away
how many would
there be left? _____

Henry threw 3 dice . . .

What did he score? _____

Joseph threw . . .

What did he score? _____

Who scored the most? _____

How many more did Joseph
need to make his score up to 15? _____

10	● AT2 level 2: can count sets of objects reliably. ● AT2 level 2: can choose appropriate operation when solving + and − problems. date:

Name . . . . Class . . . .

double
→

3	→	6
5	→	
7	→	
20	→	
	→	8
	→	18
	→	100

Use one of these numbers in each box to make these correct.

(2   5   7   8)

3 + 4 = 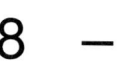   8 − = 3

4 +  = 6   × = 14

6 +  = 11   × = 10

9 − 4 =   6 × = 12

- Can double numbers.
- Mental recall of number facts.

date:

11

Name . . . . Class . . . .

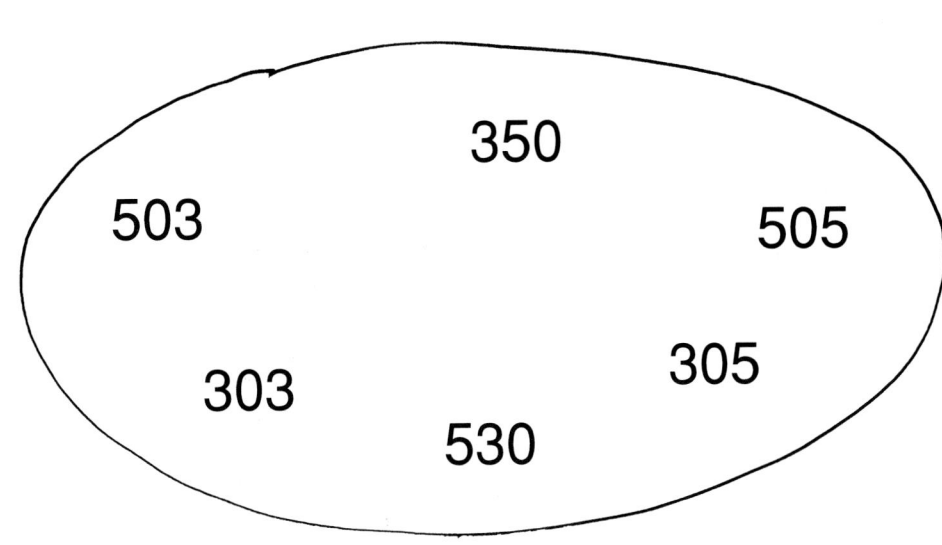

Write these numbers in order in the boxes.

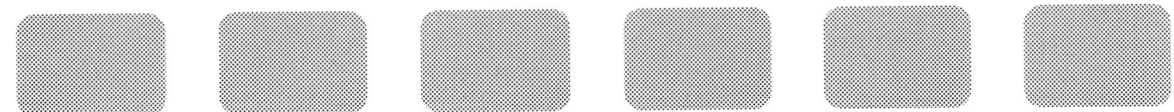

Which number is closest to 500? _____

Which number is nearest to 310? _____

Write the number that has 5 tens in it. _____

Which number is this?

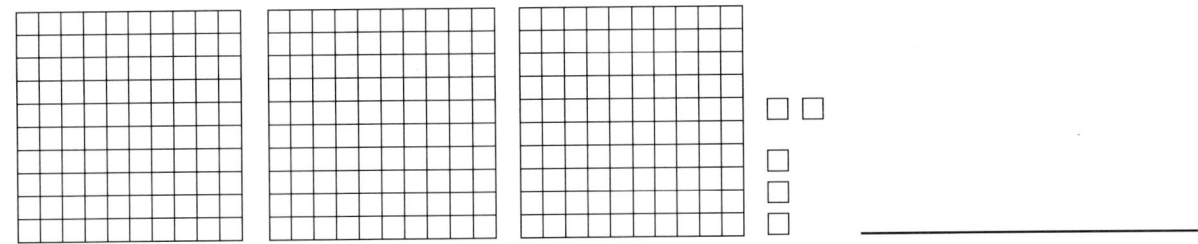

_____

| 12 | ● AT2 level 3: understands place value to 1000. |
| | ● AT2 level 3: can approximate numbers to the nearest 10 or 100. |

date:

Name . . . . Class . . . .

Fill in the missing numbers in these sequences.

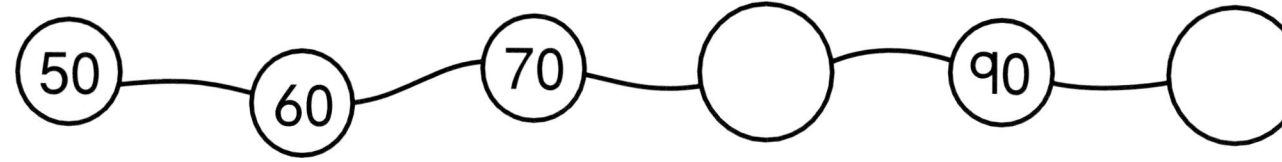

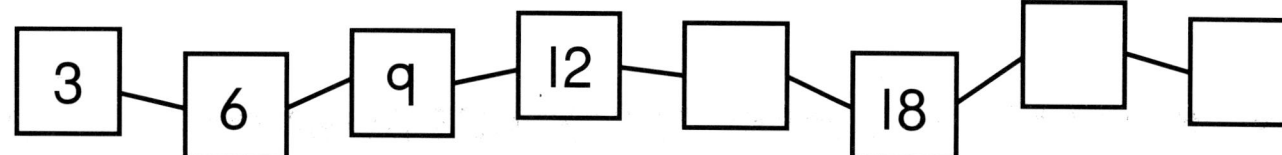

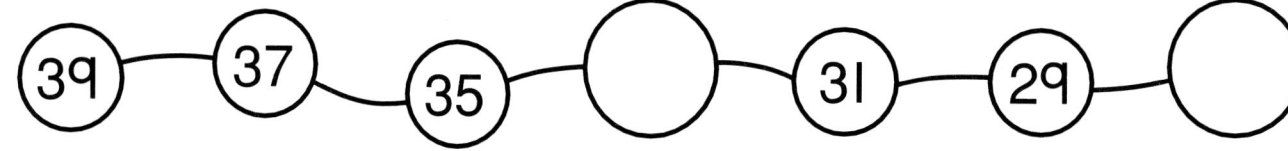

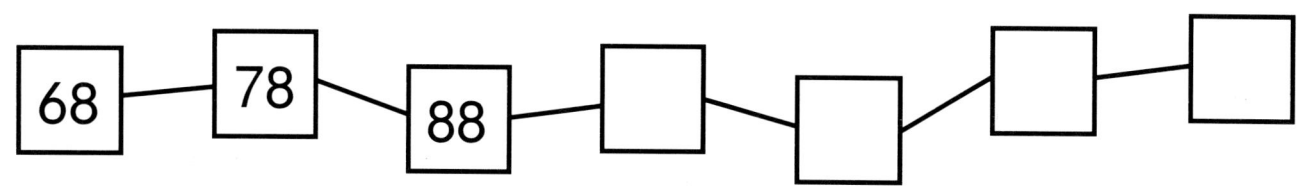

34 + 10 =

65 + 10 =

87 + 100 =

23 + 400 =

200 − 1 =

602 − 3 =

Add together 39 and 45

- AT2 level 2: recognises sequences of numbers.
- AT2 level 3: understands place value up to 1000.
- AT2 level 3: can use mental strategies to + and − with at least 2-digit numbers.

date:

13

Name . . . .   Class . . . .

Use two of these numbers to make an answer of 75.

2	150	25	4	20	3	100

☐ × ☐ = 75

☐ ÷ ☐ = 75

Complete these sequences of numbers.

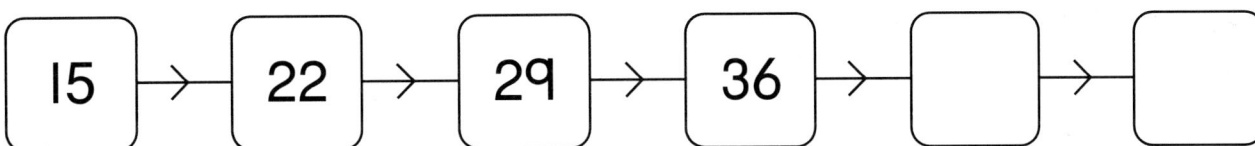

15 → 22 → 29 → 36 → ☐ → ☐

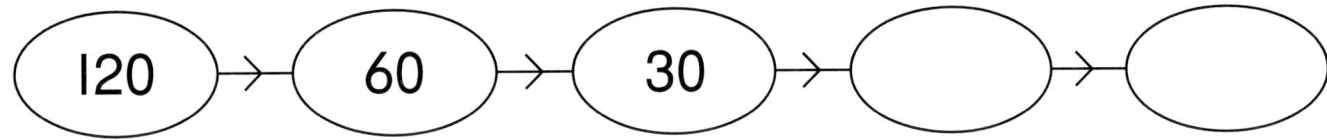

120 → 60 → 30 → ◯ → ◯

64 − 45 = ☐

92 − 18 = ☐

24 + 3 = 30 − ☐

14

- Can use × and ÷ to solve whole number problems.
- AT2 level 2: recognises sequences of numbers.
- AT2 level 3: can use mental strategies to add/subtract with 2-digit numbers.

date:

Colour half ($\frac{1}{2}$) of each of these shapes.

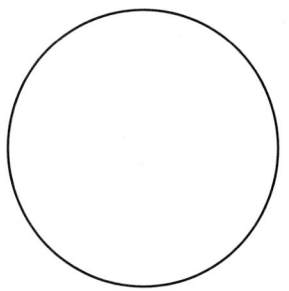  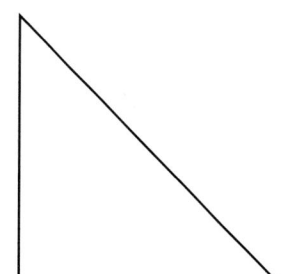 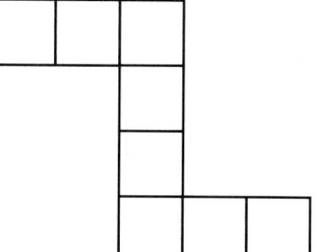

Colour a quarter ($\frac{1}{4}$) of each of these shapes.

 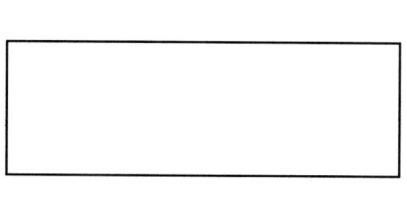 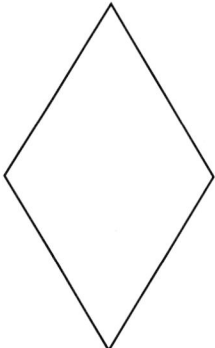

Colour $\frac{1}{2}$ of each set.

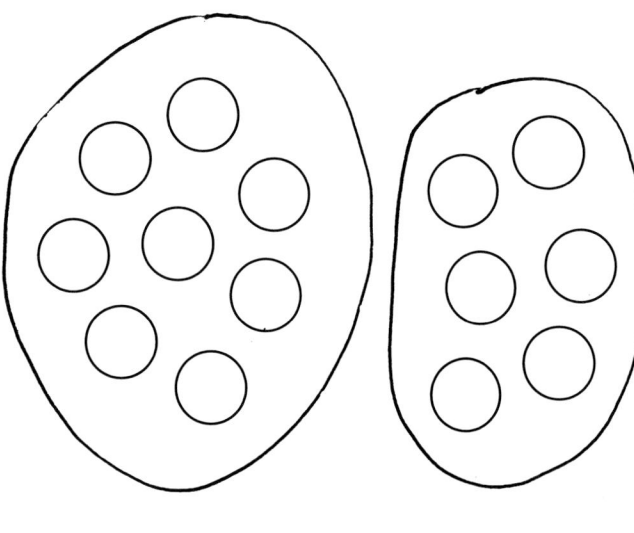 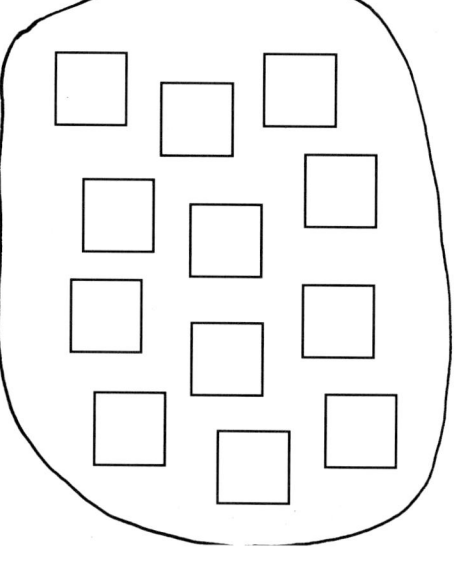 

- AT2 level 2: begins to use fractions, identifying halves and quarters.
- Can shade $\frac{1}{2}$ of a shape.
- Can shade $\frac{1}{4}$ of a shape.
- Can partition a set in 2 equal halves.

date:

15

Name . . . .   **Class** . . . .

How much? _____

How much? _____

      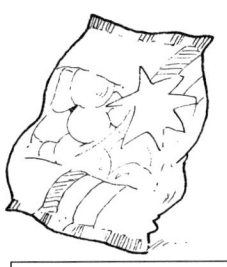   

| apple 12p | drink 15p | crisps 9p | cake 8p |

How much do these cost?

● 2 cakes _____   drink and crisps _____

● apple and cake _____   2 drinks _____

● Sarah spent 23p. Which 2 items did she buy?

_____ and _____

● Jo has 10p. He wants to buy an apple.

How much more does he need? _____

● How many apples could you buy for 50p? _____

| 16 | ● AT2 level 2: chooses appropriate operation when solving addition and subtraction problems.
 ● AT2 level 3: solves problems involving multiplication or division, including those that give rise to remainders. | ● Can total coins.
 ● Can total bills.
 ● Can solve problems involving money.
 date: |

How much? _____

How much? _____

Alice saves 50p each week.

How much has she saved after 6 weeks? _____

. . . and after 9 weeks? _____

£1.40

How many
20p coins? _____

£2.00

How many
10p coins? _____

5p

£... 45p

How many
5p coins? _____

- Can total coins.
- Can use appropriate notation to record money.
- AT2 level 3: can solve problems involving multiplication or division.

date:

17

Name . . . . **Class** . . . .

Cafe
Tea 45p

Coffee 55p

Orange Drink 20p

Iced Bun 22p

Cream Cake 39p

How much do these cost?

2 teas _____   a tea and a cream cake _____

2 coffees _____   a tea and an orange drink _____

4 orange drinks _____   a coffee and an iced bun _____

Katie has £1. How many iced buns can she buy? _____

Amin bought 2 drinks costing 75p altogether.
What did he buy?

_____ and _____

Simon had     He bought 2 cream cakes.
How much change was he given?

_____

● AT2 level 3: uses mental strategies to add and subtract numbers with at least 2 digits.
● AT2 level 3: solves problems involving multiplication/division/addition/subtraction.

18

date:

Match the words to the shapes.

cube

cuboid

cylinder

sphere

cone

Tick (✓) the shapes that will roll.

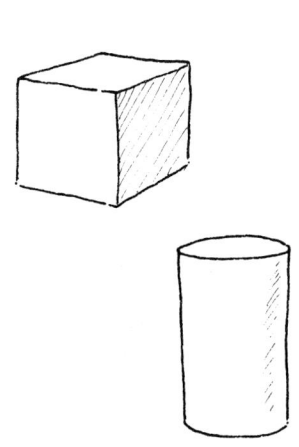

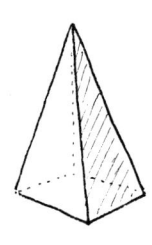

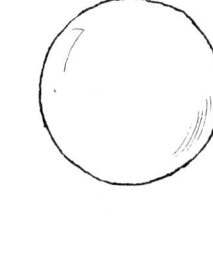

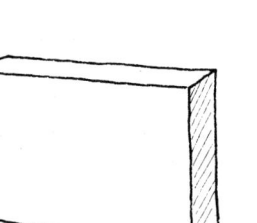

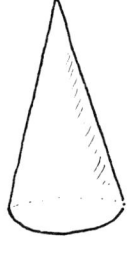

● AT3 level 2: uses mathematical names for common 3D shapes.
● AT3 level 2: describes properties of shapes.

19

date:

Name . . . . **Class** . . . .

Match the words to the shapes.

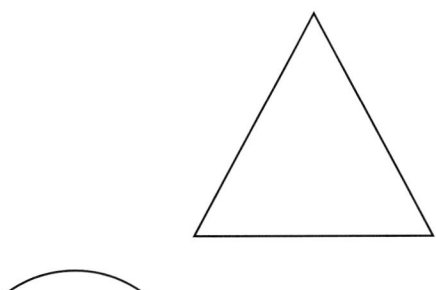

square

circle

triangle

hexagon

pentagon

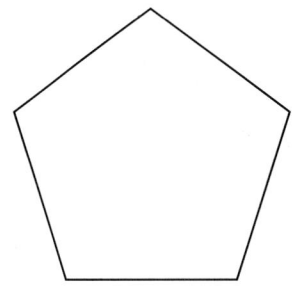

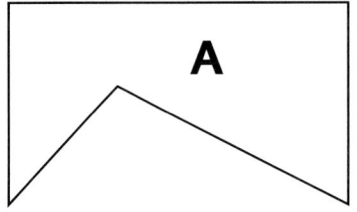

A

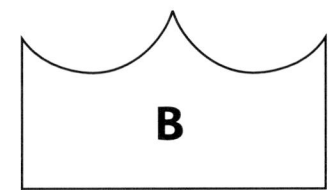

B

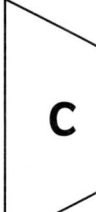

C

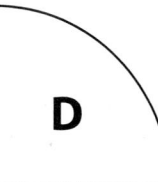

D

Which shape has 5 straight sides? _____

Which shape has no right angles? _____

Which shape has 4 corners? _____

Which shape has 2 curved sides? _____

Put a cross in one of the right angles.

20	• AT3 level 2: uses mathematical names for common 2D shapes. • AT3 level 2: describes properties of 2D shapes. • AT3 level 2: can identify right angles. date:

Here is one small square.

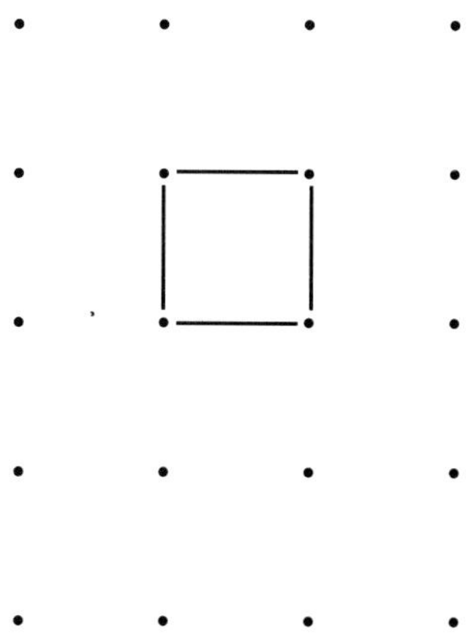

Join dots to make a rectangle
that covers 12 small squares.

▼

notes

date:

21

Name . . . . Class . . . .

Put a ring round the shapes that have reflective symmetry.
Draw the lines of symmetry on these shapes.

Draw the reflection of these shapes in the mirror line.
You may use a mirror.

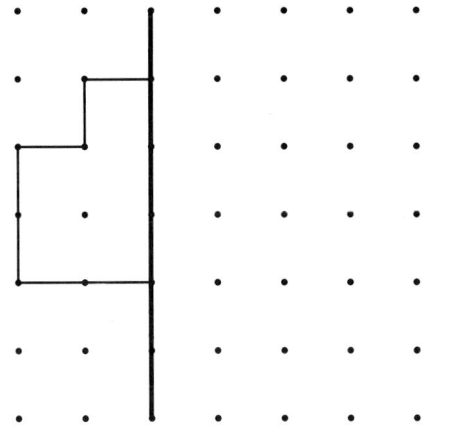
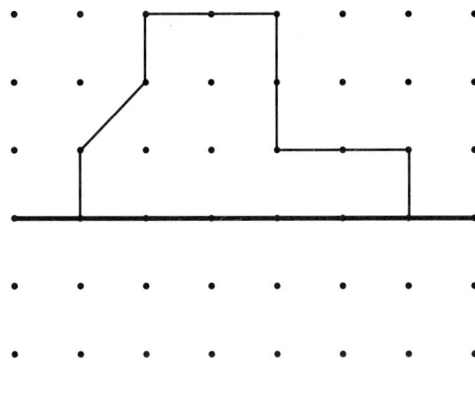

22	● AT3 level 3: uses mathematical properties such as reflective symmetry. ● Recognises reflective symmetry. ● Can draw lines of symmetry. ● Can draw mirror images of shapes. date:

Name . . . . **Class** . . . .

How heavy is the apple?

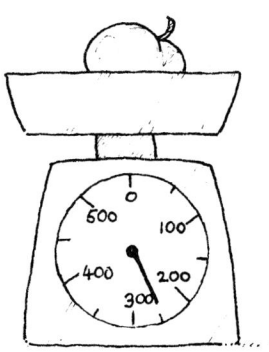

_____ grams

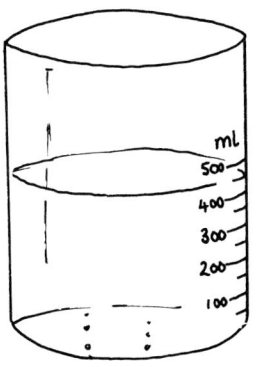

How much water is there?

_____ ml

Find the longest line. Use a ruler to measure it.

It is _____ long.

The shortest line is _____ long.

● Can use measuring instruments, interpreting numbers and scales accurately.
● Can measure accurately using a ruler.

date:

23

Name . . . . **Class** . . . .

Choose your answers from here.

4 cm

2 cm   14 cm

4 m   2 m

24 cm

2 kg

20 kg

200 g

20 g

2 l

20 l   2 ml

6 l

60 ml

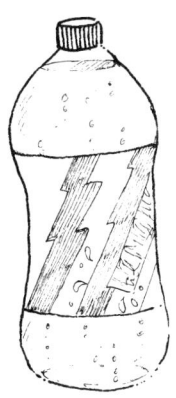

How much
lemonade? _____

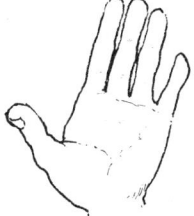

How long is
Katie's thumb?
_____

How heavy is
the banana?

_____

How long is the pen? _____

How high is this
classroom door?

_____

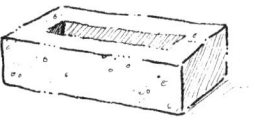

How heavy
is the brick? _____

| 24 | • AT3 level 3: uses standard units of length, capacity and mass in context.
• Can choose appropriate sets of units.
• Can make a sensible estimate. |

date:

Name . . . . **Class** . . . .

Match the clock faces to the times in the box.

7 o'clock

half past 3

2 o'clock

quarter past 6

3:00

quarter to 8

12:15

half past 11

Put a ∗ next to the clock that shows a time
that is around lunch time.

- Can read clocks: o'clock/half past/quarter past/quarter to.
- Can interpret digital time and match with analogue clock.

date:

25

Name . . . . Class . . . .

Robert woke up.

He arrived at school.

Assembly began.

Playtime began.

Playtime ended.

Maths lesson.

- When did Robert arrive at school? _____

- At what time was he having a Maths lesson? _____

- When did he wake up? _____

- He ate breakfast at 8.15.
  Put hands on the clock face to show this.

- How long did playtime last? _____

- Lunch is at half past 12.
  Put hands on the clock face to show this.

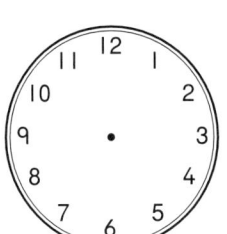

- Assembly lasted for 20 minutes.
  What was the time when it finished? _____

26

- Can read times: o'clock/half past/others.
- Can record time on analogue clocks.
- Can interpret digital time and record an analogue clock.
- Can solve problems involving passing of time.

date:

Some children used a tally chart to record the traffic passing their school.

cars 		ЖЖ ЖЖ ІІІІ
bicycles		ЖЖ ЖЖ І
pedestrians		ЖЖ ЖЖ ЖЖ ЖЖ ІІ
dogs		ІІІ

1 How many bicycles? _____

2 Were there more cars than bicycles? _____

3 How many pedestrians? _____

4 Add 2 more dogs to the chart.

   How many now? _____

5 How many more pedestrians than bicycles were there? _____

● AT4 level 3: can extract and interpret information presented in simple tables and lists.
● AT1 level 3: can use and interpret mathematical symbols and diagrams.
● AT1 PoS: can use a variety of forms of mathematical presentation.   date:

27

Name . . . . **Class** . . . .

This table shows the favourite drinks of the
children in a class.

Kind of drink	Number of children whose favourite it was.
orange	7
lemon	9
cola	7
water	3
milk	5

1 Which drink did most children choose? _____

2 Which drink did fewest children choose? _____

3 More children chose cola than milk.

   How many more? _____

4 Which drinks were chosen by the same number of children?

   _____ and _____

5 How many children are in the class? _____

28

● AT4 level 3: can extract and interpret information presented in simple tables.
● AT1 PoS: can use a variety of forms of mathematical presentation.
● AT1 PoS: can understand the language of number: most/fewest/more/same.

date:

Name . . . . **Class** . . . .

shapes where $\frac{1}{2}$ is shaded
shapes with four sides

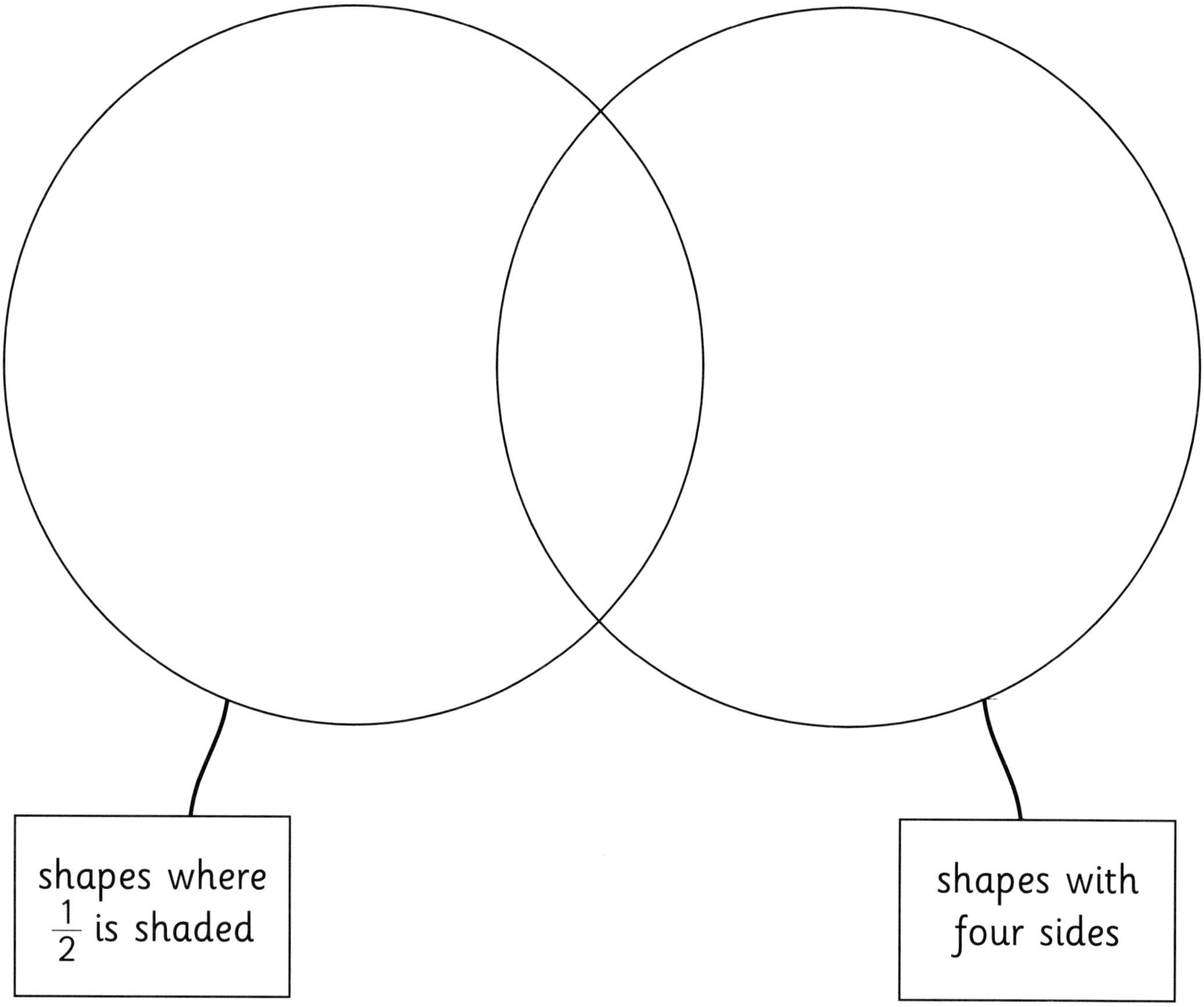

Copy the shapes below onto this Venn diagram.
Look at the labels before you decide where they should go.

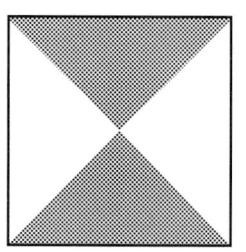

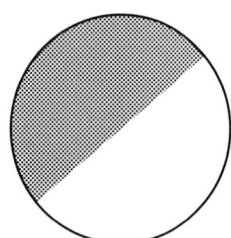

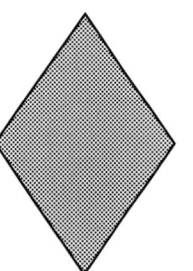

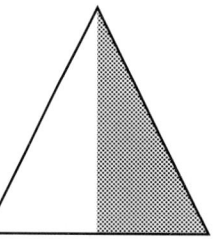

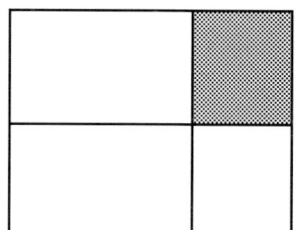

- Can understand and complete Venn diagrams.
- Can identify halves of shapes accurately.

date:

29

Name . . . . Class . . . .

	Walked to school today	did not walk to school today
age 7	Jenny Sue Kate Sam Joshua Ben	James Hayley David Pia
not age 7	Ellen Nick Ali	Laura Alex Oliver Dan George

1 How many children walked to school today? _____

2 How many 7 year olds did not walk to school? _____

3 Is Laura aged 7? _____ And Ben? _____

4 John is 6. He came to school on the bus.
   Put him in the right place on the chart.

30	● AT4 level 3: can extract and interpret information in simple tables.
	date:

Name . . . . **Class** . . . .

Class 2 made a chart showing their shoe sizes.

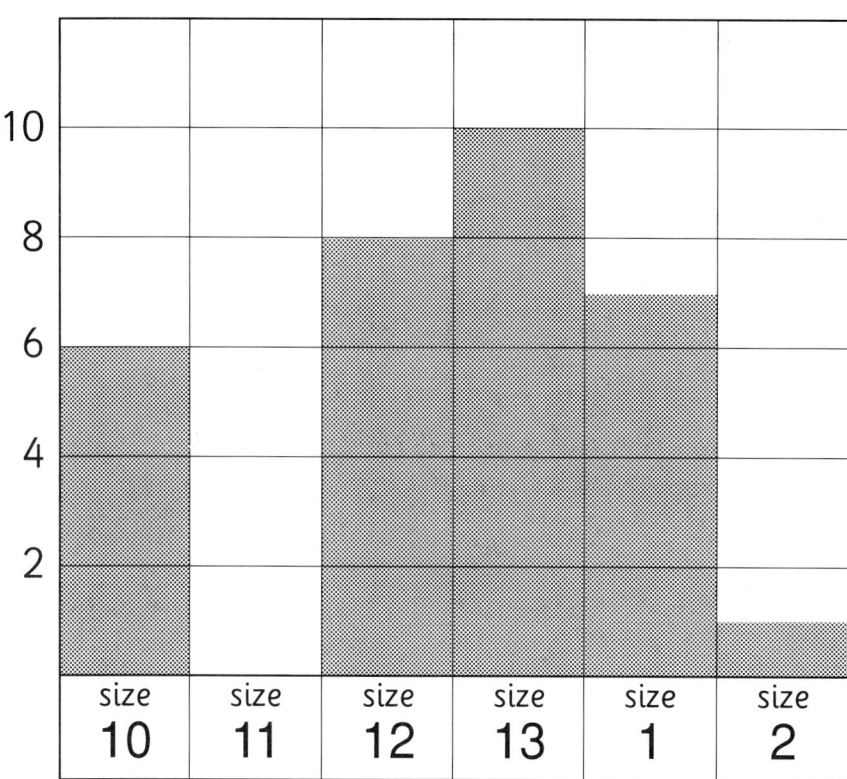

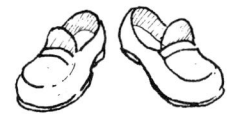

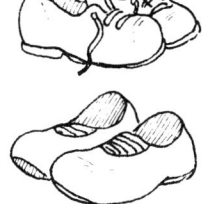

1 Which size did most children take? _____

2 Which size is not worn by anyone? _____

3 How many children take size 12? _____

4 How many children take size 1? _____

5 How many more take size 10 than size 2? _____

6 How many children are in this class? _____

● AT1 level 3: can use and interpret mathematical diagrams.
● AT1 PoS: understands the language of numbers and comparatives.
    responds to mathematical questions.
● AT4 level 3: can extract and interpret

information presented in tables. can use bar charts where the symbol represents a group of units.

date:

31